SpringerBriefs in Earth Sciences

For further volumes:
http://www.springer.com/series/8897

Rituparna Bose

Devonian Paleoenvironments of Ohio

With Foreword by Prof. David Harper
(Chairman of the International Subcommission on
Ordovician Stratigraphy and President of International
Palaeontological Association)

 Springer

Rituparna Bose
City University of New York
New York
USA

ISSN 2191-5369   ISSN 2191-5377 (electronic)
ISBN 978-3-642-34853-2   ISBN 978-3-642-34854-9 (eBook)
DOI 10.1007/978-3-642-34854-9
Springer Heidelberg New York Dordrecht London

Library of Congress Control Number: 2012953465

Printed on acid-free paper

Springer is part of Springer Science+Business Media (www.springer.com)

Parts of this monograph has been published in journal PALAIOS:

Bose, R., Schneider, C., Polly, P. D., and Yacobucci, M., 2010. Ecological interactions between Rhipidomella (Orthides, Brachiopoda) and its endoskeletobionts and predators from the Middle Devonian Dundee Formation of Ohio, United states. Palaios 25:196–208.

Parts of this monograph has been published in journal PALAIOS

This book is dedicated to Emeritus Prof. Richard D. Hoare (Department of Geology, Bowling Green State University, Ohio) who was awarded the William W. Mather Medal in 2002 by the Ohio Department of Natural Resources for his significant contributions in Ohio geology. I also want to dedicate this to my former supervisor, Prof. Margaret M. Yacobucci (Department of Geology, Bowling Green State University, Ohio) who was a constant source of inspiration in my graduate career

Foreword

It is now over 50 years since publication of Derek Ager's classic paper on the epifauna of a Devonian spiriferid yet there have been relatively few comprehensive studies on the biological and environmental aspects of these critical ecological indicators. Rituparna Bose has targeted in some detail the rich brachiopod fauna of the carbonate rocks of the Middle Devonian Dundee Formation of Ohio including taxa of the key orders of Strophomenida, Orthida, Rhynchonellida, Atrypida and Spiriferida. Some 300 specimens have been carefully examined and many, beautifully illustrated. Remarkably, epibiont activity has been almost exclusively restricted to the punctate orthide *Rhipidomella* in contrast to the distribution of such activity across a wide range of brachiopod taxa in the overlying siliciclastic facies of the Silica Formation. The taphonomy of the brachiopod shells is examined and a wide range of structures ranging from boreholes to encrustations are identified, described, and illustrated. Clearly, there are strong environmental and taphonomic controls on the scale and variation of epifaunal development in such benthic communities and Ager's studies too alerted us to the functional significance of the mode and position of borings and encrusters. This work is a rich source of data, significantly expanding our understanding of the ecological interactions between commensal taxa and their brachiopod hosts and substrates. The data and analyses presented here provide much encouragement for further research on this key area of brachiopod paleoecology.

David A. T. Harper
Professor of Palaeontology
Department of Earth Sciences
Durham University
Chairman, International Subcommission on Ordovician Stratigraphy
President of International Palaeontological Association

Preface

Exploring ancient life forms is essential to ecological and environmental studies. *Devonian Paleoenvironments of Ohio, USA* elucidates the rich paleoenvironment of the Middle Devonian sediments of Ohio in the Paleozoic era, representative of an important environment in the Michigan Basin of North America. It provides insight into how the paleoecology of extinct invertebrates living during that time can be appropriately used to reconstruct the past environment. Paleoecological interactions between brachiopods and other microinvertebrates are illustrated in detail could be with special emphasis on encrustation patterns and predatory relationships.

The book helps readers understand the various aspects of biotic relationships (mutualism, commensalism, parasitism, and predation) within an ancient ecosystem. It will be a valuable read for biologists, geologists, ecologists, environmental, and climate scientists. It may be used in under-graduate classes but will certainly help post-graduate students and advanced professionals.

The author is grateful to Prof. Margaret M. Yacobucci at Bowling Green State University, Ohio for her valuable suggestions. The foreword for the book has been written by Prof. David Harper (Chairman of International Subcommission on Ordovician Stratigraphy, 2008-present; President of International Palaeontological Association, 2006–2010).

Acknowledgments

I would like to acknowledge Margaret M. Yacobucci, Don C. Steinker and Richard Hoare from Bowling Green State University, Ohio for their kind suggestions. A special thanks to Chris Wright and David VanDeVelde for all their help.

Contents

Abstract

Epibionts seem to be more common in siliciclastic units than in carbonate units. To evaluate this difference, the paleontology of the Middle Devonian Dundee Formation has been explored. A total of 245 brachiopod specimens were collected from a fossiliferous horizon of the Dundee Formation exposed at Whitehouse Quarry and identified to the generic level. Brachiopod genera identified were Strophodonta, Rhipidomella, Rhynchotrema, Atrypa, and Mucrospirifer. All the brachiopod shells were examined under a stereomicroscope for evidence of epibionts, and preferred host taxa were determined. Epibionts are absent on all the brachiopod shells except some Rhipidomella shells. Further examination of these Rhipidomella shells under 100x magnification showed evidence of biotic interactions in 21 out of 48 specimens. Large boreholes were produced by worm borers, scars were left on a few specimens by worms, branching grooves were the traces of soft-bodied ctenostome bryozoans, and sheet-like encrustation was produced by an indeterminate group of bryozoans. Ctenostome bryozoans had a commensal relationship with their host while a few worms had a parasitic relation with the host. While one might expect encrustation on hardgrounds within this carbonate unit, field work has determined that much of the Dundee Formation was extensively bioturbated, implying a soft substrate. It may be that bioturbation mixed shells down into the substrate before epibionts could attach.

Chapter 1
Introduction

The paleontology of the Middle Devonian Dundee Formation of the Michigan Basin has not been well studied. Some sedimentological and stratigraphic information is available for this unit (Sparling 1988), but the fossils from localities such as the Whitehouse Quarry (Lucas County, northwest Ohio) have not been rigorously collected and examined. In particular, the paleoecology of this fossiliferous marine unit should be better constrained. Prior workers have shown that the strophodontids and spiriferids found in the overlying Silica Formation are often heavily encrusted. Many paleontologists have studied epifaunal growth on brachiopods of different ages and different areas, but no one has actually collected specimens from the Dundee Formation and studied them for epibionts.

Epibionts are those organisms that attach permanently to a hard substrate. Determining the interrelationships among the brachiopod hosts and different epibionts will provide evidence for commensalism, predation, parasitism, and other ecological interactions that occurred within this paleocommunity. In addition, comparison of the amount of encrustation in the carbonate Dundee Formation and the overlying siliciclastic Silica Formation will shed light on the importance of host substrates in these two different environments. Further, this study will assist future workers to compare encrustation patterns on brachiopod hosts in carbonate and siliciclastic environments.

1.1 Previous Work

1.1.1 Epibionts

Boekschoten (1966) states that biologists have gathered a lot of information on shell borers, but paleontologists and geologists have not paid much attention to the action of boring organisms and epibiontic growths. Hence, he was the first paleontologist to note that a lot of paleoecological inferences can be drawn by studying these epibionts.

R. Bose, *Devonian Paleoenvironments of Ohio*, SpringerBriefs in Earth Sciences, DOI: 10.1007/978-3-642-34854-9_1, © The Author(s) 2013

Epibionts are those organisms that attach permanently to the hard substrate of other organisms and then grow on their preferred substrates (Fig. 1.1a–d). These are often very small and/or colonial organisms; some are skeletal while others are soft-bodied. The skeletal epibionts may be branching, spiraling, or sheet-like in nature. Soft-bodied epibionts may leave traces with a dendritic pattern, straight wide grooves, or circular boring traces. The small hard skeletal organisms include corals, brachiopods, bryozoans, and crinoids ·while the soft-bodied epibionts include worm tubes, algae, and some bryozoans.

Some epibionts prefer certain hosts for encrustation. The most common host shells during the Paleozoic were brachiopods. Other hosts on which epibionts could grow include gastropods, rugose corals, bryozoans, bivalves, and crinoids. The most common epibionts in the Paleozoic are bryozoan colonies; others include worm tubes, crinoids, other brachiopods, and corals. Sparks et al. (1980) identified the most common epibionts on the fossil brachiopod host *Paraspirifer bownockeri* of the Devonian Silica Formation as *Hederella* colonies, ctenostome bryozoans, the chain coral *Aulopora microbuccinata*, and *Cornulites* worm tubes (Fig. 1.1a). In modern oceans, bivalve shells often show epibionts (Fig. 1.1b). Epibionts can even encrust mobile hosts; examples of this include living sea turtles, which can act as hosts for epibionts like barnacles, algae, and sucker fish (Fig. 1.1c), and swimming scallops heavily encrusted with barnacles and sponges (Fig. 1.1d).

By examining where the epibionts are growing, it is possible to determine the relationship between the host and the epibiont, their timing of encrustation, and the life orientation of the host shell during encrustation. Host organisms may be dead or alive at the time of attachment. If an epibiont attaches to the external surface of the host, and if it is found not covering the commissure margin, it is possible that the shell was alive during encrustation with its external surface exposed. Also, if there is evidence of healing of shell damage, it implies that the host was alive during its encrustation. If there is encrustation on the internal surface of the host, or if epibionts are found crossing the commissure, it is clear that the host was dead during encrustation.

Interrelationships between the brachiopod hosts and different epibionts provide evidence for predation, parasitism, commensalism, mutualism, and other ecological interactions. From the orientation and position of epibionts, it can be determined whether they were situated such as to derive benefit from current flow or concealment, providing further evidence that the host was living during its encrustation. Through camouflage, epibionts can mask the host's visual signals thereby inhibiting predators from attacking the host organism. In this way epibionts can benefit the host organism. On the other hand, some epibionts can harm the hosts by boring into their shells and causing severe damage. For example, some gastropods can drill holes in their hosts and feed on their soft tissues. Weakening of a shelly host by borings, such as Devonian *Atrypa* shells bored by sponges, and brachiopods from the upper Ordovician Richmond Group drilled by gastropods, are very common examples of epibionts causing harm to the host (Ager 1963).

Martindale (1992) showed that there are localized variations of encrustation within carbonate units. He found that Recent and Pleistocene Barbados reefs are

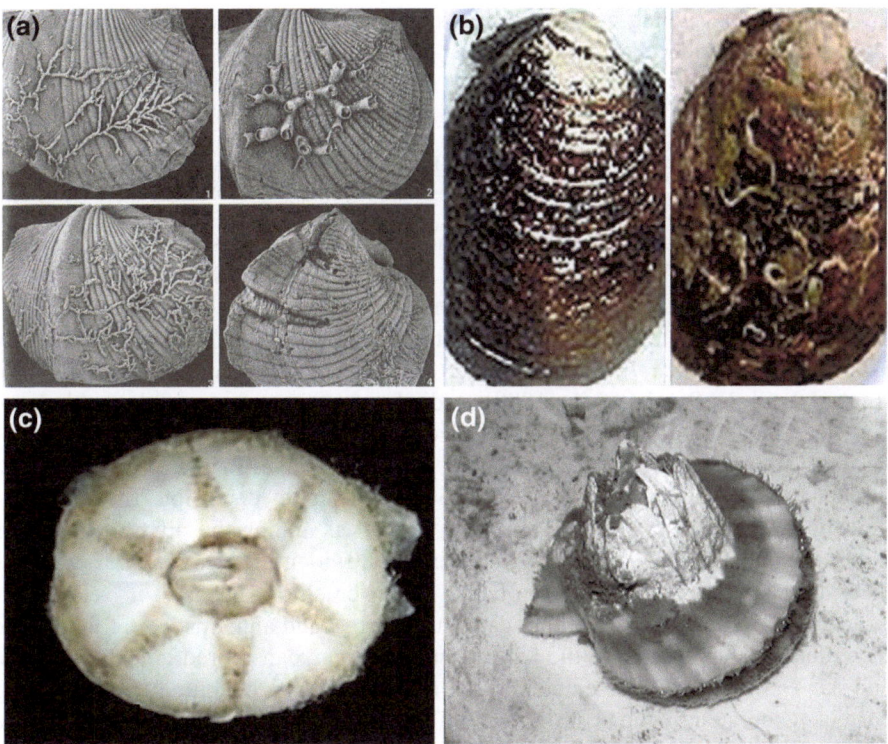

Fig. 1.1 Example of epibionts. **a** *1 Hederella* colony developed with few branches on the host brachiopod. *2* Co-occurrence of coral *Aulopora microbuccinata* and ctenostome bryozoan *Eliaspora stellatum* on the same host. *3 Hederella* confined to the brachial valve. *4 Hederella* and *Cornulites* worm tube co-existing upon the same host (from Sparks et al. 1980). **b** Worm tubes attached to the external surface of the modern bivalve (from Parsons-Hubbard 2001). **c** A sea turtle with barnacle attachments (from Pinou and Lazo-Wasem 2002). **d** A scallop with barnacle attachment (from Donovan and Bingham 2003)

encrusted by calcified epibionts like crustose coralline algae, bryozoans, foraminiferans, and serpulid worms. However, there were thick crusts of coralline algae encrusting the exposed reefs while the cryptic environments were characterized by epibionts like thin crusts of algae, bryozoans, foraminiferans, and serpulid worms. It is well known that carbonate and siliciclastic units have varied nutrient levels. Experimental results from the study sites in the Java Sea, Indonesia, have shown that shell encrustation is positively correlated with productivity (Lescinsky et al. 2002). Elevated shells from eutrophic sites like Pulau Panjang (6 m) had greater animal encrustation, and a large biovolume of animals such as mollusks and barnacles. On the other hand, shells from mesotrophic sites like Gosong Cemara (10 m), further offshore, were encrusted by coralline algae and serpulid worms, and they had a low biovolume of animals, such as bryozoans (Lescinsky et al. 2002). Thus, sites with low nutrient level in shallow marine environments have low encrustation, which can be used as models to predict the amount of encrustations in the Dundee Formation.

1.1.2 Paleontological Studies of Epibionts

Numerous workers have studied epibionts on Devonian fossils. Near the southeastern edge of the Michigan Basin, the Middle Devonian Silica Formation, the unit directly overlying the Dundee Formation, is well known for both its brachiopod fauna and their epibionts. The brachiopod community was dominated by *Paraspirifer bownockeri,* which served as host for different types of epizoans (Figs. 1.1a, 1.2). Sparks et al. (1980) studied the epifauna preserved on this large host, the distribution of each kind of epizoan, and also determined which valve of the host had more epizoans attached to it, which helped in interpreting the life orientation of the brachiopod host. They also tried to determine the interrelationships among host shell and epizoans as well as among the different epizoans that settled on the same host. Kesling et al. (1980) identified 38 kinds of epizoans that selected the hard surface of the large brachiopod host for settlement. Bryozoans, corals, other brachiopods, echinoderms, and annelids were found to be the epibionts that settled on the large host. Some epizoans bored holes in the host and kept themselves protected, while other epizoans benefited from feeding currents generated by the host or attached in an area where they could eat away the soft tissues of the host.

Epibionts were found to be abundant on the brachial valve, from which it was interpreted that the host probably rested on the pedicle valve during its life, with its brachial valve exposed. Then, by examining the possible position of a host and an epibiont, life relationships were classified into different categories, some of which are symbiosis (mutualism and commensalism), toleration, or antagonism (antibiosis, exploitation—parasitism and predation). Interrelationships among colonial corals and the host were mutualistic in that both benefited from their relation. The host seemed to be protected by the stinging cells of the coral colony, while at the same time the coral also received anchorage for its colony. On the other hand, the *Cornulites* worm tube and *Clionides,* the boring sponge, had a parasitic relationship with the host in that the *Cornulites* intercepted food from the feeding currents of the host and *Clionides* attacked the anterior edge of the host shell, causing damage and curtailing growth for a while, although the damage did not seem to be fatal. Co-occurrences of different epizoans on the same host were also determined (Fig. 1.2). Some epibionts that were present on the internal surface of the shell or those that crossed the commissure line are believed to have been attached after the death of the host. So, life-death associations were also clearly determined (Kesling et al. 1980).

Hoare and Walden (1983) did further research work on the borings found in the brachiopod host *Paraspirifer bownockeri* collected from the Middle Devonian Silica Formation of northwestern Ohio. The borings tended to be straight to slightly curved, following the ornamentation of the host shell, and about 59 % of the borings covered the brachial valves. The borings were identified to be those of polychaete worms that were responsible for cessation of the growth of the host or causing damage to their shell. These deformations were then seen to be occupied by *Cornulites* worms.

The Norway Point Formation in the Devonian Traverse Group of Michigan contains a single species of brachiopod, *Spinocyrtia clintoni* (Pitrat and Rogers 1978). The authors studied 280 specimens and about 70 % of these specimens were found to bear epibionts like *Spirorbis* species, *Cornulites* species, *Paleschara* species, *Hederella* species, and *Aulocystis commensalis*. Cameron (1969) identified the Middle Devonian fossilized circular shell borings from the Marcellus Formation near Morrisville, New York, as those of polychaete worms. Brachiopod-coral (*Mucrospirifer-Aulopora*) associations are known from the Middle Devonian Upper Yuchiang Formation of Kuangsi Province in China (Shou-Hsin 1959). Symbiotic associations of a Devonian spiriferid, *Spinocyrtia iowensis*, from the Upper Devonian Cedar Valley Limestone of Iowa and adjacent states were also common (Ager 1960). Encrustation and borings were common on brachiopods and horn corals from the Upper Devonian Lime Creek Formation of Rockford, Iowa (Anderson and Megivern 1982).

Schneider and Webb (2004) found differences in the rate of encrustation on Devonian brachiopods and Mississippian brachiopods across the Frasnian-Famennian extinction. The rate of encrustation was higher in the Devonian ribbed

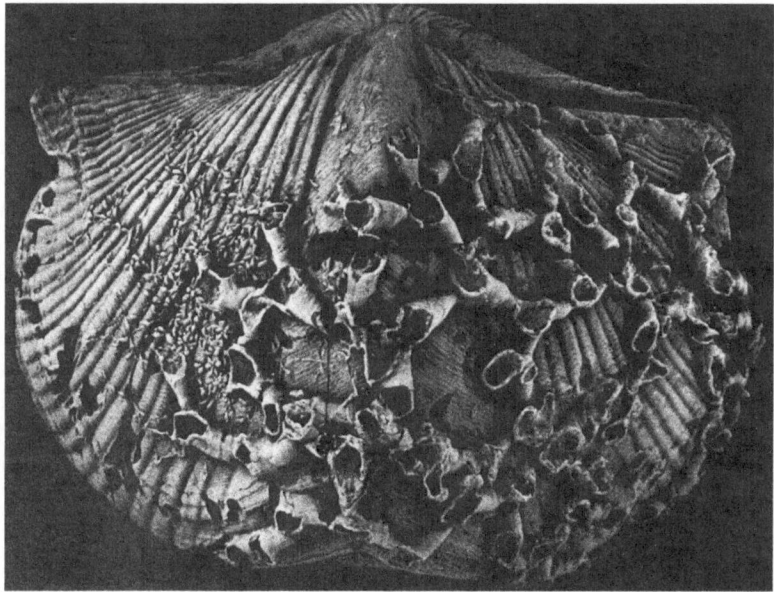

Fig. 1.2 Epibionts on *Paraspirifer bownockeri* from the Middle Devonian of Ohio (from Sparks et al. 1980)

brachiopods compared to the smooth-shelled brachiopods, while in the Missis-sippian, smooth-shelled brachiopods were preferentially encrusted. Plicate and costate brachiopods like atrypids and strophodonts were lost across the Frasnian-Famennian extinction boundary, and there was an increase in abundance of smooth-shelled and spinose brachiopods in the Mississippian. At the Geological Society of America meeting in 2004, during the discussion of this talk, Carlton Brett (University of Cincinnati) noted that, in New York State, the same bra-chiopods would be encrusted if in Hamilton Group shales (stratigraphically equivalent to Ohio's Silica Formation) but not encrusted if in the Onondaga Limestone (equivalent to the Dundee Formation). No one in attendance knew why this difference exists (Yacobucci, personal communication, 2004). It is therefore scientifically very interesting to see if the brachiopods encrusted in the Dundee Formation were the same as in the overlying Silica Formation and if encrustation is much rarer in the Dundee Formation.

Studies of epibionts on brachiopod hosts in carbonate units both older and younger than the Devonian are relevant to the current study. *Cornulites* worms were reported to have encrusted brachiopods, bryozoans, and a trilobite from the Upper Ordovician Waynesville Formation of southwestern Ohio (Morris and Rollins 1971). Hoare (2003) studied the brachiopod fossils of the Mississippian (Carboniferous) Maxville Limestone in Ohio. He examined collected specimens from Ohio and Indiana Universities as well as some of his own collections. He first identified 14 different taxa of brachiopod host and then examined the epifauna. A dozen or more *Spirorbis* worm tubes were found to be attached to a single host shell with their openings towards the shell margin. Another worm tube, *Cornulites,* was found attached to two different hosts with the same orientation. Some bry-ozoans were also found to be encrusted as multiple colonies on one host specimen. Peterson and Hoare (1973) studied the occurrences of the epizoan rugose corals on host brachiopods *Composita subtilita* and *Neospirifer dunbari* from the Pennsyl-vanian Ames Limestone (Conemaugh Formation) of Ohio. The common epibionts attached to brachiopods and crinoid columnals include barnacles, bryozoans, corals, and echinoderms. Eleven specimens of a coral-brachiopod association were collected from two localities in Guernsey County, Ohio, and examined to deter-mine the causes of the specific position of the epibionts and the life orientation of the brachiopod host.

Epibionts on fossil crabs are also known from the late Middle Danian Lime-stones at Fakse Quarry, Denmark (Jakobsen and Feldmann 2004). Epibiont cov-erage on modern mollusk shells transported into a high-energy beach environment derived from the carbonate reef and lagoon systems in the northeastern Caribbean was found to be very low due to shells being in constant motion (Parsons-Hubbard 2005). It is noteworthy from previous work that brachiopods in limestones of other ages can show clear evidence of encrustation.

References

Ager DV (1960) The epifauna of a Devonian spiriferid. Q J Geol Soc Lond CXVII(465):1–10

Ager DV (1963) Principles of paleoecology. McGraw-Hill, New York 371 p

Anderson WI, Megivern KD (1982) Epibionts from the Cerro Gordo member of the Lime Creek formation (Upper Devonian), Rockford, Iowa. Proc Iowa Acad Sci 89(2):71–80

Boekschoten GJ (1966) Shell borings of sessile epibiontic organisms as palaeoecological guides (with examples from the Dutch coast). Palaeogeography, Palaeoclimatology, Palaeoecology 2:333–379

Cameron B (1969) Paleozoic shell boring annelids and their trace fossils. Zoologist 9:689–703

Donovan DA, Bingham BM (2003) Effects of barnacle encrustation on the swimming behaviour energetics, morphometry and drag of the scallop *Chamys hastata*. <http://fire.biol.wwu.edu/donovan/donovan.html>. Retrieved 5 June 2006

Hoare RD (2003) Brachiopods from the Maxville Limestone (Mississippian) of Ohio. Ohio Division of Geological Survey, Report of Investigations, No. 147, pp 1–16

Hoare RD, Walden RL (1983) *Vermiforichnus* (Polychaeta) borings in *Paraspirifer bowneckeri* (Brachiopoda: Devonian). Ohio J Sci 83(3):114–119

Jakobsen SL, Feldmann RM (2004) Epibionts on *Dromiopsis rugosa* (Decapoda: Brachyura) from the Late Middle Danian Limestones at Fakse Quarry, Denmark: novel preparation techniques yield amazing results. J Paleontol 78(5):953–960

Kesling RV, Hoare RD, Sparks DK (1980) Epizoans of the Middle Devonian brachiopod *Paraspirifer bownockeri*: their relationships to one another and to their host. J Paleontol 54(6):1141–1154

Lescinsky HL, Edinger E, Risk MJ (2002) Mollusc shell Encrustation and Bioerosion rates in a modern Epeiric Sea: taphonomy Experiments in the Java Sea, Indonesia. Palaios 17:171–191

Martindale W (1992) Calcified epibionts as palaeoecological tools; examples from the recent and pleistocene reefs of barbados. Coral Reefs 11(3):167–177

Morris RW, Rollins HB (1971) Palaeoecologic interpretation of *Cornulites* in the Waynesville formation (Upper Ordovician) of southwestern Ohio. Ohio J Sci 71(3):159–170

Parsons-Hubbard K (2001) Sea skeletal remains on the slope and shelf. <http://www.at-sea.org/missions/skeletal/day2.html>. Accessed 5 June 2006

Parsons-Hubbard K (2005) Molluscan taphofacies in recent carbonate reef/lagoon systems and their application to sub-fossil samples from reef cores. Palaios 20:175–191

Peterson RM, Hoare RD (1973) Epizoan rugose coelenterates from the Ames Limestone (Conemaugh) of Ohio. The Compass 50(3):22–24

Pinou T, Lazo-Wasem E (2002) Turtle epibiont project (The Collections). <http://www.yale.edu/peabody/collections/iz/iz_epibiont.html>. Accessed 5 June 2006

Pitrat CW, Rogers FS (1978) *Spinocyrtia* and its epibionts in the traverse group (Devonian) of Michigan. J Paleontol 52(6):1315–1324

Schneider CL, Webb A (2004) Where have all the encrusters gone? Encrusting organisms on Devonian versus Mississippian brachiopods. Geological Society of America Abstracts with Programs, vol 36 (5), p 111

Shou-Hsin C (1959) Note on the paleoecological relation between Aulopora and Mucrospirifer. Acta Paleontol Sinica 7:502–504

Sparks DK, Hoare RD, Kesling RV (1980) Epizoans on the brachiopod *Paraspirifer bowneckeri* Stewart) from the Middle Devonian of Ohio. University of Michigan Museum of Paleontology, Papers on Paleontology, No. 23, pp 1–50

Sparling DR (1988) Middle Devonian stratigraphy and conodont biostratigraphy, north-central Ohio. Ohio J Sci 88(1):2–18

Chapter 2
Research Goals

Previous workers have observed that epibionts are common in siliciclastic units (Sparks et al. 1980). The main purpose of my research is to see whether the Devonian brachiopods of the carbonate Dundee Formation in the Whitehouse Quarry have less encrustation than the overlying Silica Formation, a siliciclastic unit. I will also determine if the same kinds of brachiopods were encrusted in these two units.

Overall, my work will be the first to:

(a) examine brachiopod specimens in this rock unit for encrustations,
(b) investigate the ecological interactions within the fauna of the Dundee Formation,
(c) test possible causes for the low rate of encrustation seen in carbonate units.

Reference

Sparks DK, Hoare RD, Kesling RV (1980) Epizoans on the brachiopod *Paraspirifer bowneckeri* (Stewart) from the Middle Devonian of Ohio. University of Michigan Museum of Paleontology, Papers on Paleontology, No. 23, pp 1–50

Chapter 3
Geological Background

The Dundee Formation is located on the flanks of the Michigan Basin, and my study area is located in the southeastern part of the Michigan Basin at Whitehouse, Lucas County, Ohio (Figs. 3.1, 3.2). The Dundee Formation is situated on the western flank of the Findlay Arch, which was a topographic high during the Devonian. This topographic high separated and isolated the Michigan Basin from the Appalachian Basin (Fig. 3.2). The Chatham sag between the Findlay and Algonquin Arches may have acted as a depression for trapping the lower Dundee sediments (Fig. 3.2; Birchard and Risk 1990). The Dundee Formation formed during the Middle Devonian, about 391–380 million years ago. More specifically, the Eifelian-aged Dundee Formation underlies the Givetian-aged fossiliferous Silica Formation of the Traverse Group (Fig. 3.3).

Detailed core and outcrop studies of the Middle Devonian Dundee Limestones of southwestern Ontario have been performed by Birchard and Risk (1990). The Dundee Formation exposed in southwestern Ontario is composed of many 6–12 cm grayish-blue limestone beds. In some sections it is partially dolomitized, with chert nodules in the lower part. Six informal units were identified (Table 3.1) in this formation that helped in interpreting the stratigraphy and sedimentation of the Dundee limestones accumulated in the Michigan Basin (Birchard and Risk 1990). Three lithofacies (facies 2, 4, and 5) recognized within the Dundee strata were common to both the Michigan and Appalachian Basins. Birchard and Risk (1990) found that the two basins responded to sea level fluctuations very differently for the uppermost Dundee strata. The depositional environment for the uppermost Dundee strata in the Appalachian Basin represents a final transgressive episode with deposition of deep-water argillaceous mudstones, while that for the uppermost Dundee strata in the Michigan Basin was described as a sea-level stillstand, with episodes of reworking and winnowing. Repetitions of lagoonal and open-shelf facies in Appalachian Basin regions, firm ground development, and coarse, reworked packstone to grainstone pulses give evidence of sea-level fluctuations in middle Dundee time. The presence of brachiopods, bryozoans, rare crinoids, and corals in limestones represents a shallow-shelf environment (Birchard and Risk 1990).

R. Bose, *Devonian Paleoenvironments of Ohio*, SpringerBriefs in Earth Sciences,
DOI: 10.1007/978-3-642-34854-9_3, © The Author(s) 2013

The Dundee Formation in the Whitehouse Quarry at Lucas County, Ohio, underlies the Silica Formation stratigraphically and it has been correlated with the limestones of other regions (Fig. 3.3). Based on conodont data, the Eifelian-aged Dundee Formation in northwestern Ohio is faunally related to the basal Delaware Limestone and Upper Columbus Limestone of central Ohio and Upper Onondaga Limestone of western New York (Sparling 1988).

The stratigraphy of the Dundee Formation at Whitehouse has recently been re-studied by Wright (2006). A composite measured section produced by Wright is shown in Fig. 3.4. Detailed unit descriptions have been compiled in Table 3.2. The

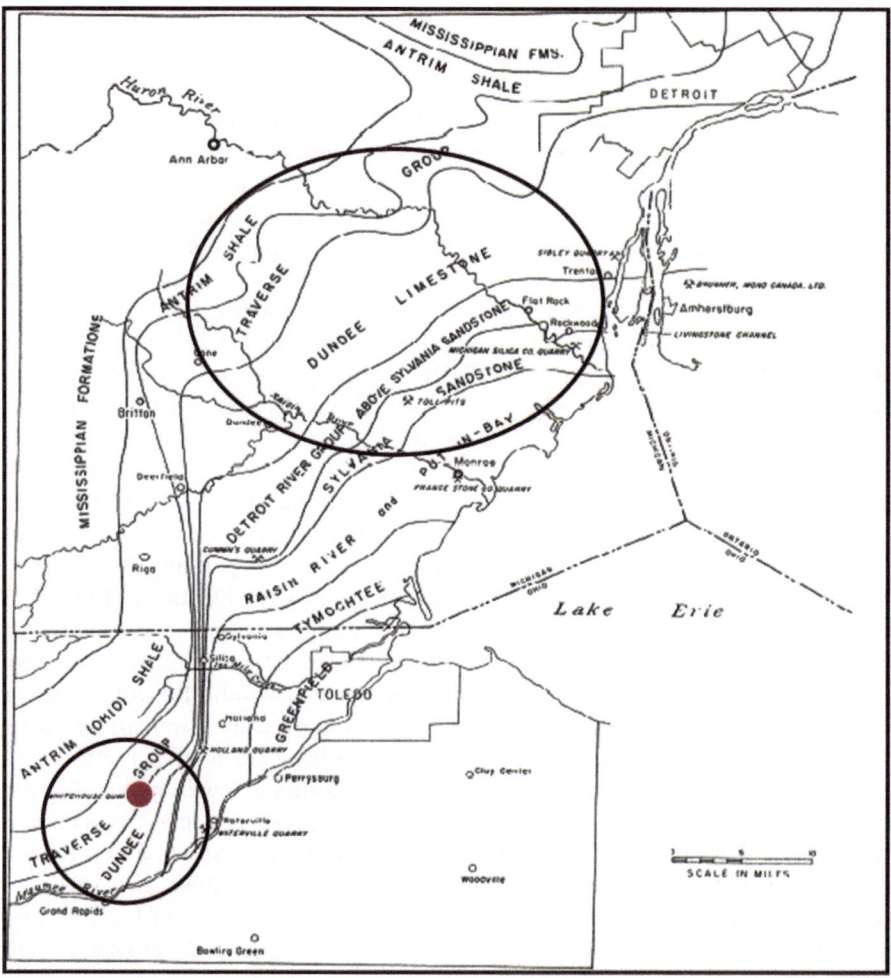

Fig. 3.1 Geologic map of southeastern Michigan and northwestern Ohio, showing the narrow belt of Dundee Formation along the Lucas County Monocline. *Black dot* marks the location of Whitehouse Quarry, Whitehouse, Lucas County, Ohio (modified from Ehlers et al. 1951)

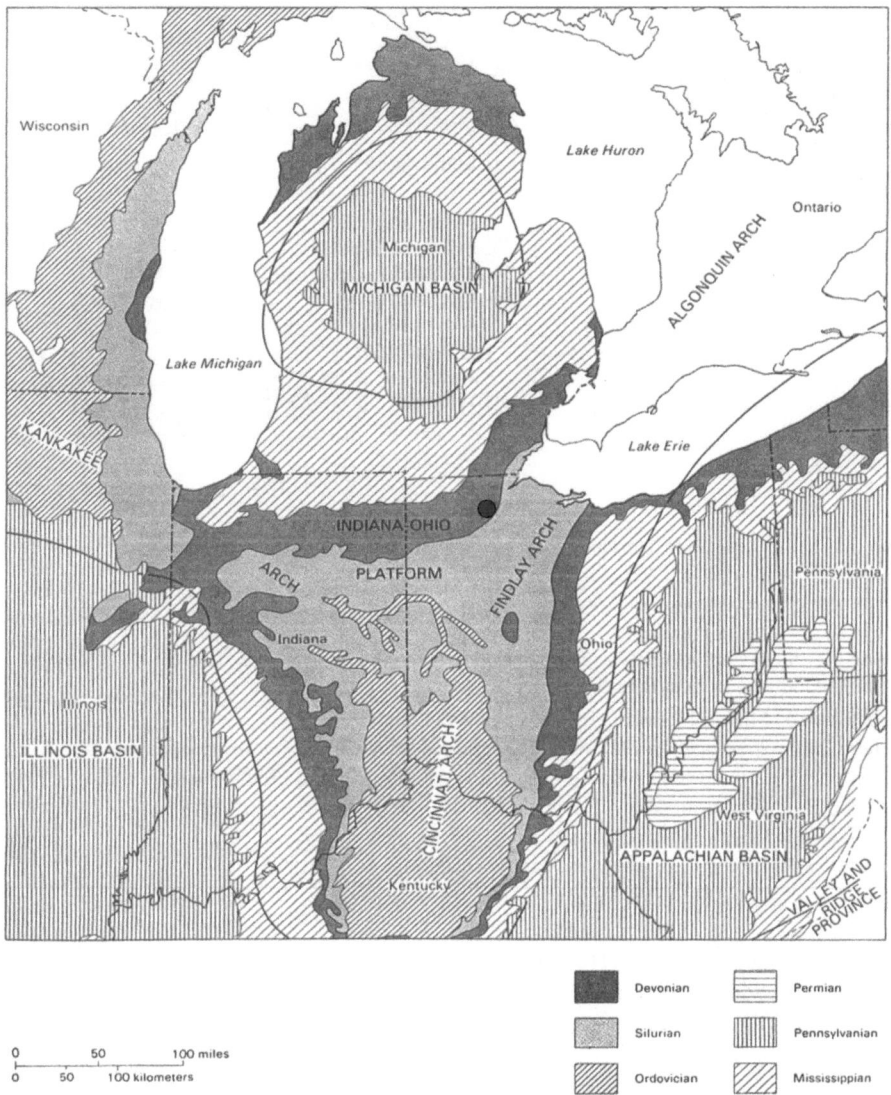

Fig. 3.2 Regional geologic structures of Ohio and adjacent states. *Black dot* marks location of Whitehouse Quarry (modified from Carlson 1991)

Dundee Formation exposed in the Whitehouse Quarry, Lucas County, Ohio, is about 8.7 m in thickness, composed of grayish-blue limestones, with evidence of partial dolomitization. Limestone beds are cherty in the lower part of the exposed Dundee Formation. The beds sampled for this study included units 10, 11, and part of unit 12 (Wright 2006). These units differed in lithology from each other and were in total 0.32 m thick. The lower unit was a packstone, 0.19 m in thickness,

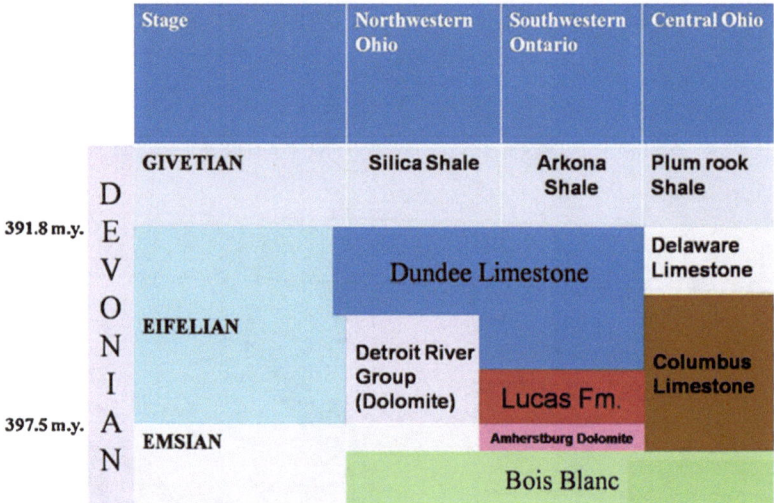

Fig. 3.3 Conodont-based correlation of the Middle Devonian strata in the Lake Erie region (modified from Sparling 1988)

Table 3.1 Stratigraphic section of the Middle Devonian Dundee Formation from southwestern Ontario (modified from Birchard and Risk 1990)

Lithofacies 6: Muddy bioturbated wacke-packestone unit
Lithofacies 5: Argillaceous, bioturbated brachiopod mudstone to wackestone facies
Lithofacies 4: Crinoid-brachiopod grainstone to wackestone firm-ground facies
Lithofacies 3: Cherty mudstone facies
Lithofacies 2: Cherty bioclastic facies
Lithofacies 1: Bioturbated, dolomitic sandy wackestones

grading upward into wackestone of about 0.07 m in thickness, finally grading into a nodular bedded grainstone, 0.06 m in thickness.

3.1 Paleoenvironmental Interpretation of the Dundee Formation (Whitehouse Quarry)

Packstones and grainstones are grain-supported limestones that represent high-energy environments. Wackestones are mud-supported limestones with more than 10 % grains that represent quieter to moderate energy level conditions. Mudstones represent deposition in a low-energy environment and quiet-water conditions. The lowest Dundee unit in this quarry contains wackestone grading upward into 3.0 m of mudstone, representing a deepening upward sequence, which implies there was a transgression. Then there was a shallowing upward sequence with high-energy

Fig. 3.4 Stratigraphic
section of Whitehouse Quarry
(modified from Wright 2006).
Numbers mark the units listed
in Table 3.2. *Small arrow*
marks the sampled beds

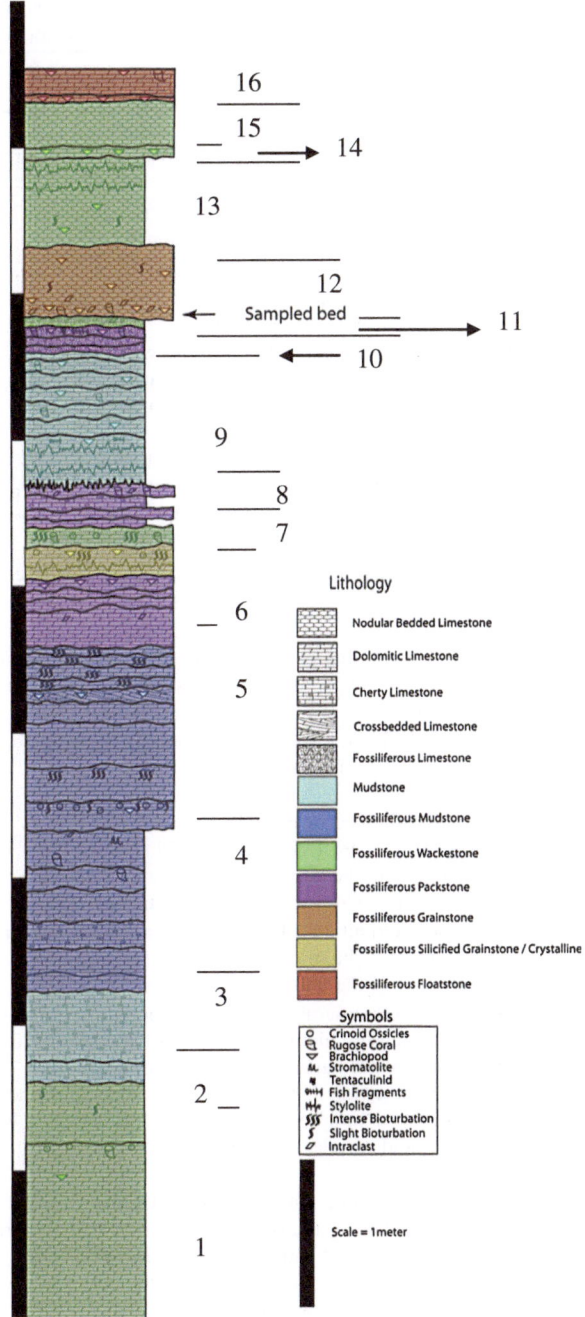

Table 3.2 Description of Dundee Formation exposed in the Whitehouse Quarry, Lucas County, northwestern Ohio region (with reference to Fig. 3.4; from Wright 2006)

Unit	Thickness (m)	Cumulative thickness (m)
16. Fossiliferous dolomitic floatstone with abundant brachiopods in the lowermost and uppermost part of the unit; rare rugose corals in the lowermost part	0.22	8.71
15. Nodular bedded wackestone with no fossil evidence	0.30	8.49
14. Dolomitic wackestone with abundant brachiopods	0.07	8.19
13. Fossiliferous nodular bedded wackestone with stylolites at the top, and the lower part slightly bioturbated and characterized by a few brachiopods	0.63	8.12
12. Fossiliferous nodular bedded grainstone with abundant brachiopods, rugose corals and intraclasts in the lowermost part of the unit with evidence of slight bioturbation	0.48	7.49
11. Fossiliferous wackestone with rare intraclasts	0.07	7.01
10. Fossiliferous packstone with diverse fossil assemblage of brachiopods, rugose corals, tentaculids and fish fragments	0.19	6.94
9. Dolomitic mudstone with stylolites at the lower part; brachiopods and rugose corals distributed in the overall unit	0.93	6.75
stylolite contact		
8. Fossiliferous dolomitic packstone with cherty packstone intercalations; intraclasts and rugose corals present in the upper part	0.30	5.82
7. Fossiliferous dolomitic wackestone, intensely bioturbated, characterized by presence of crinoid ossicles and brachiopods	0.15	5.52
6. Fossiliferous silicified or crystalline grainstone, slightly dolomitized, characterized by crinoid ossicles and brachiopods in the upper part of the unit with intense bioturbation; stylolites present at the bottom	0.22	5.37
5. Fossiliferous dolomitic packstone characterized by numerous brachiopods in the upper part, and intraclasts in the middle part of the unit	0.48	5.15
4. Fossiliferous dolomitic mudstone with uppermost 0.3 m of the unit intensely bioturbated, next 0.1 m cross-bedded mudstone, lower part of the next 0.7 m dolomitic mudstone intensely bioturbated and upper part of the lowermost 0.2 m dolomitic mudstone characterized by many crinoid ossicles, rare brachiopods and slight bioturbation	1.26	4.67
3. Fossiliferous dolomitic mudstone intercalated with fossiliferous cherty mudstone in the lower middle part of the unit; uppermost part of the dolomitic mudstone characterized by stromatolites, and intraclasts and rugose corals in the upper part of the unit	1.15	3.41
2. Cherty mudstone with no fossil assemblage	0.63	2.26

(continued)

Table 3.2   (continued)

Unit	Thickness (m)	Cumulative thickness (m)
1. Fossiliferous dolomitic wackestone with diverse fossil assemblage of crinoid ossicles, rare rugose corals, and brachiopods in the upper 0.6 m of the unit, with evidence of slight bioturbation at the top	1.63	1.63
Total thickness	8.7	

conditions represented by deposition of packstone and grainstones (units 5–8). The deposition of a 0.93-m thick mudstone above these coarser units show there was another transgressive cycle and low-energy conditions. The deposition of several layers of wackestone followed, with intercalations of packstone and nodular bedded grainstone representing high-energy conditions and the onset of a regressive cycle. Bioturbation, chert replacement, and mud intraclasts are characteristics of packstones and grainstones that are common throughout the section, suggesting a very shallow, subtidal depositional environment. The presence of nodular bedded structures and stylolites provides evidence for chemical compaction. Thus, the Dundee Formation in Whitehouse Quarry is characterized by trangressive-regressive cycles,

Fig. 3.5 A fossiliferous horizon of the Dundee Formation in the Whitehouse Quarry, Lucas County, Ohio. *Black circles* show the units from which the fossils were collected

Fig. 3.6 T-shaped branching network of burrows in units 10–12 of the Dundee Formation identified as *Thalassinoides*

with fossil assemblages representing deposition in a subtidal, shallow marine, carbonate environment.

The Dundee Formation in the Whitehouse Quarry was marked by numerous fossils toward the top of the unit and fewer near the bottom (Fig. 3.4). Fragmented and abraded shells suggest that the horizon from which the fossils were collected for study was a storm bed. It is possible that strong storm currents stirred dead shells from above storm wave base and caused their transportation and re-deposition. Reid (1994) determined the nature and significance of hummocky cross-stratification and associated storm features in the Devonian Dundee Formation of Grand Rapids, Ohio (a part of the Dundee Formation that is older than the units exposed in Whitehouse Quarry). This gives further evidence that there was some storm influence in the Devonian Dundee Formation.

The Dundee environment around 350 million years ago was a shallow, tropical, high-energy environment with occasional large-scale storms. The Dundee Formation was deposited below normal wave base but above storm wave base. There is evidence that the Middle Devonian sea was never very deep, but was below storm wave base at times. The sediments of this age in most parts of the Midwest denote epineritic-biostromal environments with abundant corals and brachiopods (Cooper 1957). The sediments in the Middle Devonian Dundee Formation of

Fig. 3.7 Y-shaped branching burrows in units 10–12 of the Dundee Formation identified as *Thalassinoides*

southwestern Ontario mainly represent a shallow shelf environment with deposition near or below storm wave base with abundant brachiopod, coral, and crinoid fragments. There is evidence of transgressive cycles interrupted by minor sea-level fluctuations in these sediments (Birchard and Risk 1990). These local observations are consistent with more regional paleoenvironmental interpretations. Most of Ohio was dry land during early Devonian time, although the sea still covered eastern Ohio. Ohio was in equatorial latitudes. In Middle Devonian time, warm, shallow seas deposited limy sediments. In late Devonian time, the Ohio sea became stagnant; circulation was poor, and the water was generally anoxic. Thick layers of black, organic-rich, uranium-bearing mud were deposited in these "stinking seas" (Hansen 1999).

3.2 Previous Paleontological Faunal Lists from the Dundee Formation

The name Dundee Limestone was first used for the rocks previously exposed in the abandoned Pulver Quarry in Dundee, Monroe County, Michigan (Ehlers et al. 1951). The Whitehouse Quarry is east of the town of Whitehouse in Ohio, worked

Fig. 3.8 Criss-cross pattern of the trace fossils in units 10–12 of the Dundee Formation

by the Whitehouse Stone Company in the early 20th century. The Dundee Limestone of Whitehouse was characterized by numerous fossils collected and listed by Stauffer (1909) and Bassett (1935). Their combined faunal lists for the Dundee Formation are as follows:

Anthozoa—*Cladopora, Favosites, Zaphrentis*
Brachiopoda—*Amphigenia, Athyris, Atrypa, Camarotoechia, Chonetes, Cryptonella, Cyrtina, Eunella, Glyptodesma, Leptostrophia, Orthothetes, Pentamerella, Pholidops, Pholidostrophia, Productella, Rhipidomella, Schizophoria, Spirifer, Stropheodonta* (=*Strophodonta*)
Bryozoa—*Cystodictya, Fenestella*
Bivalvia and Rostroconchia—*Aviculopecten, Paracyclus, Pterinea, Conocardium*
Cephalopoda—*Gyroceras, Orthoceras*
Gastropoda—*Callonema, Loxonoma, Murchisonia, Platyceras, Pleurotomaria*
Conularida—*Cadeolus, Pentaculites*
Stromatoporoidea—*Stromatopora, Syringostroma*
Tentacuitoidea—*Tentaculites*
Trilobita—*Dipterus, Phacops, Proetus*
Charophytes—*Calcisphaera*
Worm tubes—*Coleolus*

Fig. 3.9 Extensive bioturbation in the Dundee Formation in the same bed as Fig. 3.8

The Dundee Formation exposed at Sibley Quarry, about two miles north of Trenton, Wayne County, Michigan, was characterized by *Favosites, Atrypa, Brevispirifer, Cyrtina, Rhipidomella, Stropheodonta,* and *Paracyclus,* as listed by Ehlers et al. (1951).

3.3 Trace Fossils in the Dundee Formation

The empty dark circles in Fig. 3.5 mark the level of the fossiliferous units 10, 11, and 12 of the Dundee Formation where some interesting burrows were photographed (Figs. 3.6, 3.7, 3.8, 3.9, 3.10). There was no direct boring or encrustation of bedding surfaces seen in the Dundee Formation, which suggests that the substrate was not a cemented hardground. Further, trace fossils observed in this formation, identified as *Thalassinoides,* were also abundant in units 6 and 7.

Thalassinoides burrows are known to have formed either in softgrounds, categorized within the *Cruziana* ichnofacies, or firmgrounds, grouped under the *Glossifungites* ichnofacies (Ekdale et al. 1984). The intense level of bioturbation (Figs. 3.9, 3.10) also noticed in units 10, 11, and part of 12, from which brachiopods were collected for this study, suggests that these were softgrounds.

Fig. 3.10 Additional extensive bioturbation in the same bed as Fig. 3.8

Trace fossils were more abundant in units 11 and 12 as compared to unit 10 of the burrowing zone. Rare body fossils like crinoid ossicles, tentaculites, and a few brachiopods were observed in the lowermost unit of the burrowing zone whereas brachiopods were most abundant in units 11 and 12. Thus, in the Dundee Formation, these trace fossil associations seem to have been formed in a subtidal shallow marine environment below fair weather wave base and above storm wave base.

Overall, the Dundee Formation represents a shallow marine carbonate substrate with evidence of intense bioturbation, suggesting these were carbonate soft-grounds. From the moderate size (0.5–1.0 cm in diameter) of the burrows, it seemed they formed in a well-oxygenated environment. The water depth seemed to be moderate (below a fair wave base but above a storm wave base), with some storm influence.

References

Basset CF (1935) Stratigraphy and paleontology of the Dundee Limestone of southeastern Michigan. Geol Soc Am Bull 46:425–462
Birchard MC, Risk MJ (1990) Stratigraphy of the middle Devonian dundee formation, southwestern Ontario. Ontario Geological Survey Miscellaneous Paper No. 150, pp 71–86

Carlson EH (1991) Minerals of Ohio. Ohio Div Geol Surv Bull 69:155

Cooper GA (1957) Paleoecology of middle Devonian of eastern and central United States. Geol Soc Am Mem 67:249–278

Ehlers GM, Stumm EC, Kesling RV (1951) Devonian rocks of southeastern Michigan and northwestern Ohio. Fieldguide for Geological Society of America, Detroit meeting, p 40

Ekdale AA, Bromley RG, Pemberton SG (1984) Ichnology: Trace Fossils In Sedimentology And Stratigraphy. SEPM Short Course No. 15, Society for sedimentary geology, Tulsa, p 317

Hansen MC (1999) America's volcanic past, Ohio (The Geology of Ohio—The Devonian, Ohio Geology). <http://vulcan.wr.usgs.gov/LivingWith/VolcanicPast/Places/volcanic_past_ohio.htm>. Accessed 6 June 2006

Reid WH (1994) Nature and significance of hummocky cross-stratification and associated storm features in the Devonian Dundee Formation, Grand Rapids, Ohio. Unpublished Masters Thesis, Bowling Green State University, Bowling Green, p 98

Sparling DR (1988) Middle Devonian stratigraphy and conodont biostratigraphy, north-central Ohio. Ohio J Sci 88(1):2–18

Stauffer CR (1909) The middle Devonian of Ohio. Ohio Geol Surv Bull 10:204p

Wright CE (2006) Paleoecological and paleoenvironmental analysis of the Middle Devonian Dundee Formation at Whitehouse, Lucas County, Ohio. Unpublished Master's Thesis, Bowling Green State University, Bowling Green, p 195

Chapter 4
Materials and Methods

Fieldwork was conducted in the Whitehouse Quarry in Lucas County, Ohio, about 14 miles northwest of Bowling Green (Fig. 3.3). A total of 245 specimens of brachiopod shells were collected by bulk sampling from fossiliferous units 10, 11, and the lower part of 12 (Fig. 3.6) near the top of the Dundee Formation (Fig. 3.7, Table 3.2). This package is about 0.32 m thick, with a lower packstone 0.19 m in thickness grading upward into wackestone of about 0.07 m in thickness, finally grading into a nodular bedded grainstone. Brachiopod shells were collected throughout the thickness of the bed, particularly from the packstone and the lower part of grainstone. This horizon chosen for study was extremely fossiliferous, with high fossil abundance and diversity. The presence of abundant intraclasts in this horizon indicates that these were storm beds. Moreover, the fossils collected from this bed were mostly a transported death assemblage, evident from signs of mechanical pre-burial damage and the presence of fragmented shells. The fossils collected for this study were mostly partially recrystallized, thus making epibiont study more challenging.

Further, a few non-brachiopod epibiont hosts, a snail specimen, and a rugose coral specimen were collected by Chris Wright from the same horizon that I used for epibiont study. Bioturbation exists both in the sampled fossiliferous horizon, that is, the units 10, 11, and 12, as well as in units 6 and 7, 1.30–1.65 m below this sampled fossiliferous horizon.

Ten loose samples of brachiopod shells and 72 slabs with brachiopod shells partially embedded in them were examined under a stereomicroscope using 40x magnification. Each brachiopod shell was identified to the generic level based on its morphological characters. Under 40x magnification, only *Rhipidomella* shells showed some evidence of encrustation. Therefore, the 48 specimens of *Rhipidomella*, with and without surface evidence of borings and encrustations, were selected and examined under 100x magnification. Those with some evidence of surface encrustation and borings were photographed using a high resolution SPOT Insight digital camera attached to the microscope. The nature and position of epibionts were noted, along with their identification as to genus. The shape, nature, and position of the boreholes were also determined. The diameter of boreholes and

R. Bose, *Devonian Paleoenvironments of Ohio*, SpringerBriefs in Earth Sciences, DOI: 10.1007/978-3-642-34854-9_4, © The Author(s) 2013 dimensions of branching grooves were determined using SPOT image analysis software. Further, all *Rhipidomella* shells with encrustations and borings were counted separately and the percentage of hosts with different biological trace types was graphed. The relationship between the host and the epibiont was interpreted from the position of the epibionts on the host brachiopod specimen. The type and position of epibionts on *Rhipidomella* host brachiopods of the limestone unit were compared to those on *Rhipidomella* shells of the overlying Silica Formation. The frequency of encrustation of the Dundee Formation was also compared to that of the Silica Formation (Sparks et al. 1980). In addition, the ichnologic character of the formation was also investigated in order to interpret aspects of the depositional environment of this unit relevant to the degree of encrustation.

Reference

Sparks DK, Hoare RD, Kesling RV (1980) Epizoans on the brachiopod *Paraspirifer bowneckeri* (Stewart) from the Middle Devonian of Ohio. University of Michigan Museum of Paleontology, Papers on Paleontology, No. 23, pp 1–50

Chapter 5
Results: Paleontological Analysis

Out of 245 brachiopod specimens counted from the samples, 163 were identified as *Strophodonta*, 48 as *Rhipidomella*, 17 as *Rhynchotrema*, 9 as *Atrypa,* and 8 as *Mucrospirifer* (Table 5.1). The higher taxonomic groups that these brachiopod genera belong to are as follows:

Strophodonta—Order Strophomenida
Rhipidomella—Order Orthida
Rhynchotrema—Order Rhynchonellida
Atrypa—Order Spiriferida
Mucrospirifer—Order Spiriferida

5.1 Taphonomy of Shells

Specimens of *Strophodonta*, *Rhynchotrema*, *Atrypa,* and *Mucrospirifer* are all poorly preserved. *Strophodonta* shells show evidence of recrystallization. Internal surfaces of these strophodonts are preserved with prominent cardinal processes. *Rhynchotrema* shells are partially embedded in rock matrix with evidence of replacement as the mode of preservation. *Atrypa* shells are preserved as external molds while *Mucrospirifer* shells have both valves preserved with evidence of replacement as the mode of preservation. *Rhipidomella* shells are also poorly preserved with part of the shells embedded in rock matrix. Many *Rhipidomella* shells were recrystallized and some shells had punctae filled with pyrite crystals (Figs. 5.8, 5.9). Some *Rhipidomella* hosts show evidence of mechanical breakage (abrasion and fragmentation) (Figs. 5.26a–b).

Many *Rhipidomella* shells provide evidence for mechanical post-burial damage (compaction) (Figs. 5.23, 5.25a–b, 5.27b).

Only one valve of each shell was available for study, as these were more or less partially embedded in the rock. It was difficult to determine which valve was preserved, the ventral or the dorsal, as two valves were not exposed for study. Also, since two valves were not available for study, it was hard to determine the

R. Bose, *Devonian Paleoenvironments of Ohio*, SpringerBriefs in Earth Sciences,
DOI: 10.1007/978-3-642-34854-9_5, © The Author(s) 2013

Table 5.1 Brachiopod taxa sampled

Brachiopod genus	Number of shells	Percentage (%)
Strophodonta	163	66.53
Rhipidomella	48	19.59
Rhynchotrema	17	6.93
Atrypa	9	3.67
Mucrospirifer	8	3.26
Total	245	100.0

time of encrustation, boring, or any other biological activity on the host shell. The pinkish-white colored *Strophodonta* shells did not show evidence of any biological activity. Some shells were characterized by a unique spotty texture that seemed like sheet-like encrustation at first sight, but they were actually microstructures on the internal surface of the shell. Thus, it was found that most of the strophodont shells have their internal surfaces preserved with their prominent cardinal processes. *Atrypa, Mucrospirifer,* and *Rhynchotrema* shells also showed no biological associations.

Out of all the brachiopod genera, only *Rhipidomella* shells (Fig. 5.1) show some evidence of biological association. Twenty-one specimens out of 48 potential hosts, that is, 44 % of the *Rhipidomella* shells, show some evidence of encrustation, boring, branching grooves, and scarring (Table 5.2). The other 56 % of the shells show no such evidence. Some shells had numerous small holes following the ornamentation pattern of radial ribs; some had branching grooves that could be produced by borers. Some shells were characterized by large, shallow, or deep scars and some shells were compressed in certain places with evidence of scars. A few shells were characterized by sheet-like covers with a spotty texture that could be either encrustation or just the microstructure of the shell, with only the inner surface of the shell preserved (Table 5.2, Fig. 5.6).

5.2 Trace Types

5.2.1 Sheet-Like Spotty Encrustation

Flat, thin sheets covering the external shell surface, thus hiding the ornamentation of the shell have been termed sheet-like encrustations.

Table 5.2 Types of biological traces on *Rhipidomella* shells

Types of biological traces	Number of shells	Percentage of shells (%)
Large boreholes	1	4.76
Scars (tapered in some)	2 probable, 2 possible, 2 inconclusive	28.57
Branching grooves	13	61.94
Sheet-like encrustation	1	4.76
Total	21	100.0

Fig. 5.1 A whole shell of
Rhipidomella with boreholes
along radial ribs of the shell.
This figure also gives a close
view of what this brachiopod
genus looks like

5.2.2 Branching Grooves

Traces of straight, curved, deep or shallow, narrow or wide bifurcating grooves
ranging in diameter from 0.04 to 0.20 mm are left by some organisms on the
external shell surface.

5.2.3 Scars

A scar refers to damage caused to the shell while growing (either due to
mechanical breakage or due to a predator or parasite) that affects the later growth
of the shell, leading to an irregular deformed shell.

5.2.4 Large Boreholes

Holes of diameter 0.16–0.28 mm made by some borers on the external shell
surface are defined as large boreholes.

Evidence for more than one type of organismal interaction on the same host
shell was observed on two different specimens (Figs. 5.12, 5.15). For instance, the
trace of a wide groove has been observed in shells with sheet-like encrustation
(Fig. 5.12).

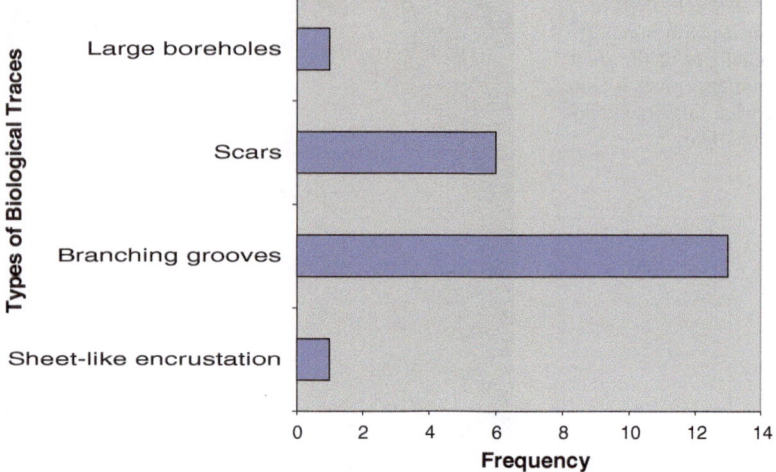

Fig. 5.2 Frequency of *Rhipidomella* host shells showing characteristic evidence of different types of biological activity

5.3 Nature and Position of Traces on Rhipidomella Shells

One *Rhipidomella* shell showed evidence of sheet-like encrustation, 13 shells have branching grooves, 6 shells show scars, and only 1 shows some evidence of large boreholes (Table 5.2, Fig. 5.2).

5.3.1 Microstructures

Brachiopods have various microstructures, which could be mistaken for epibiont traces (Figs. 5.3–5.5). Shells of all living brachiopods except rhynchonellids are pierced by slender cylindroids less than 1 micrometer (0.001 mm) in diameter called canals or very large chambers up to 20 µm (0.02 mm) or more in diameter defined as punctae (Williams et al. 1997), which respectively accommodate the

Fig. 5.3 Shell structure of a brachiopod penetrated from inside by large, regularly arranged, elongated perforations called punctae, normal to the surface (modified from Clarkson 1986, p. 42)

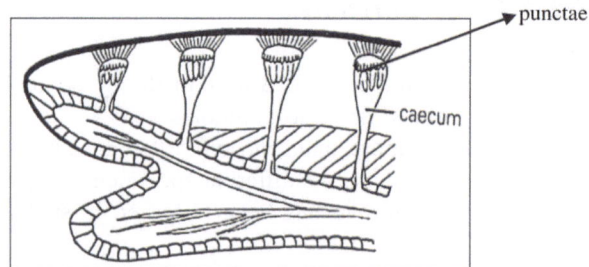

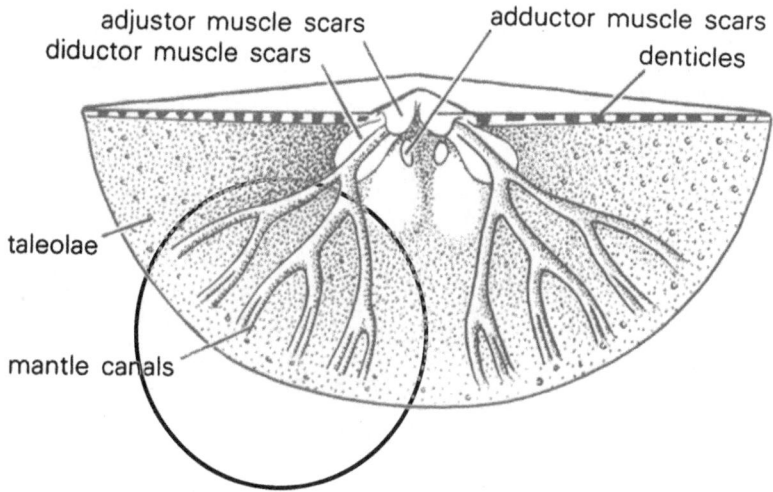

Fig. 5.4 Internal view of the pedicle valve of a brachiopod showing the mantle canal system (modified from Clarkson 1986, p. 136)

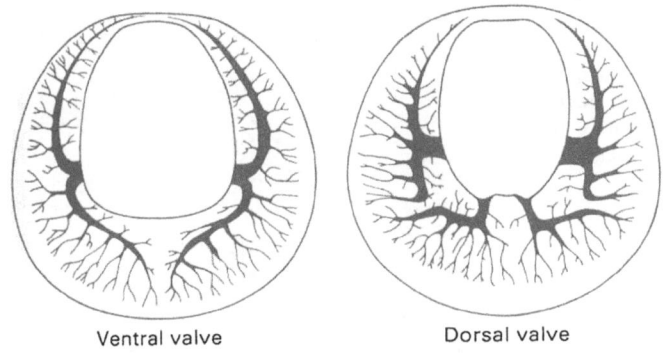

Ventral valve   Dorsal valve

Fig. 5.5 Mantle canal patterns in an inarticulate brachiopod (modified from Rowell and Grant 1987, p. 463)

membrane-bound secretions of the outer epithelium or tubular papillose out-growths of the mantle, called caeca (Fig. 5.3). In other words, punctae are elongated perforations, normal to the surface penetrated from the inside of the shell, less than 0.02 mm in diameter (Fig. 5.3). Orthids can be both punctate or impunctate (Rowell and Grant 1987). The *Rhipidomella* shells studied for epibionts in this research were observed to be punctate. These punctae could be mistaken for predatory borings and so the diameter of the cavities in these specimens were measured and found to be similar to those of punctae. Caeca within the punctae are known to be filled with protein and lipids, thereby causing predatory borers to avoid these shells. But this does not mean that punctate shells are not

Fig. 5.6 *Rhipidomella*
specimen 26C. Punctae
following the coarse radial
ribs of a *Rhipidomella* shell.
Black arrows point to the
punctae (*Note* Software uses
the term "Length" for all
linear measurements)

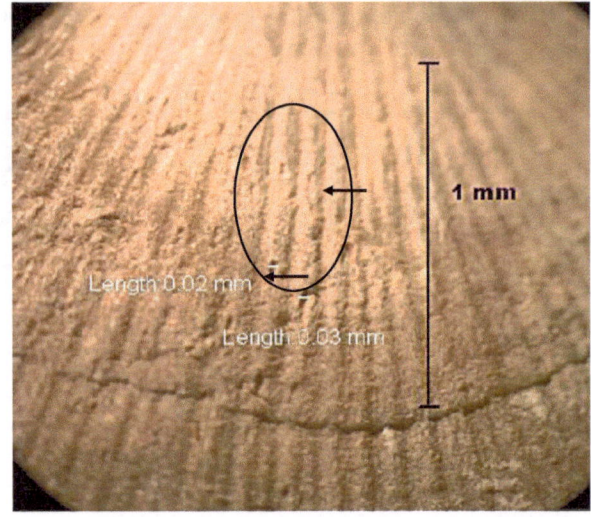

Fig. 5.7 *Rhipidomella*
specimen 16. Numerous
elongated punctae along
radial ribs crossing each
successive earlier shell
margin of the host shell
(10x). *Black arrows* point to
the elongated punctae

bored; the rate of boring in the punctate shells could be the same as in impunctate
shells (Rowell and Grant 1987).

Numerous punctae have been observed within 10 *Rhipidomella* shells that have
been examined for epibionts. These punctae vary in their shape and position on
these host specimens. For instance, some punctae of diameter 0.01–0.04 mm are
noted along the radial ribs in the central region (Fig. 5.6) and in some host
specimens there are numerous elongated punctae along radial ribs crossing each
successive earlier shell margin of the host shell (Fig. 5.7).

Fig. 5.8 *Rhipidomella*
specimen 24. Small punctae
filled with pyrite (25x). *Black
arrow* points to the punctae
filled with pyrite

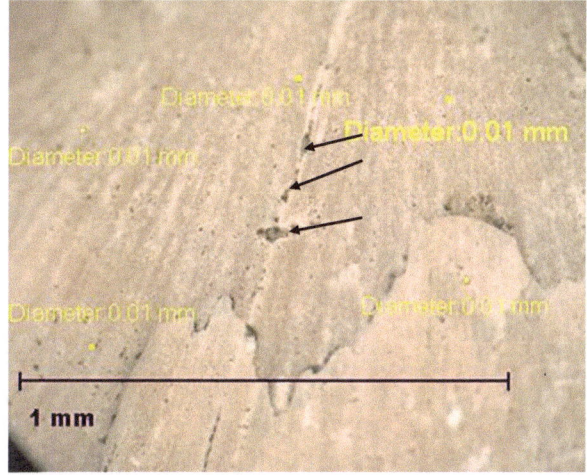

Fig. 5.9 *Rhipidomella*
specimen 12. Punctae along
ornamented radial ribs with
some filled tiny holes (16x).
Black arrows point to the
punctae along radial ribs

Another specimen had a few punctae of diameter 0.01 mm filled in by pyrite (Figs. 5.8, 5.9). One specimen was observed to have tapered scars in the central portion along with a few elongated punctae (Fig. 5.10). These punctae could be easily mistaken for holes created by some epibiont organism, but their diameter and position suggest that these were just the punctae. Mantle canals are less than 0.001 mm in diameter (Figs. 5.4, 5.5). Branching grooves observed in some specimens could be just the microstructures of the shell (the mantle canal patterns). However, their diameter (ranging from 0.04 to 0.15 mm) was found to be larger than that of the mantle canals, which proved that these were more likely traces of some epibionts.

Fig. 5.10 *Rhipidomella*
specimen 16. *Black circle*
shows the tapered scars in the
central portion and a few
elongated punctae (10x)

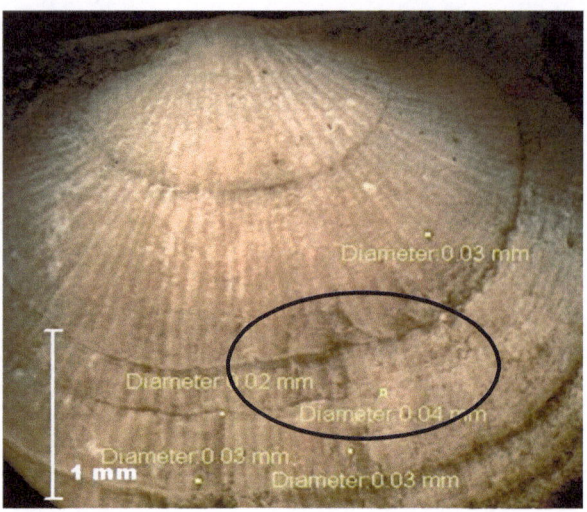

5.3.2 Sheet-Like Spotty Encrustations

Some *Rhipidomella* hosts were characterized by a unique sheet-like spotty texture, with minute spots of diameter approximately 0.005 mm, that looks like encrustation by bryozoans (Fig. 5.11).

These could be merely aspects of the brachiopod shell's microstructures preserved and not any encrusting bryozoan attached to the host valve. Outer layers of these specimens seem to have been broken off, so it is quite obvious that these were only microstructures. Just one specimen (Fig. 5.12) shows a bryozoan sheet-like

Fig. 5.11 *Rhipidomella*
specimen 23. Microstructures
underneath the external shell
surface, which has been
broken (20x)

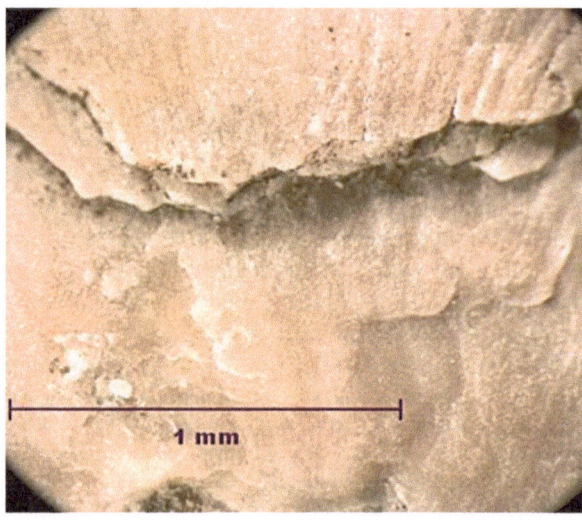

Fig. 5.12 *Rhipidomella*
specimen 19. Spotty
encrustation along the left
lateral margin close to the
commissure with the trace of
a wide groove (20x). Note
that "Length" here refers to
the width of the groove

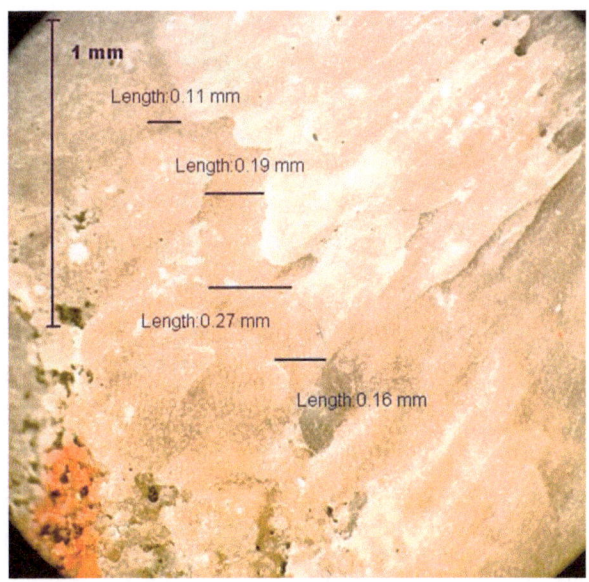

encrustation with a wide groove 0.1–0.2 mm in diameter separating the sheet into
two parts. This sheet with larger prominent spots (similar to bryozoan colonies)
seems to blanket the external surface of the shell along the left lateral margin close to
the commissure. This structure is not present underneath the broken shell surface.
Hence, these may be bryozoan encrusters and not microstructures. The wide groove
seems to be the trace of a worm boring.

5.3.3 Branching Grooves

Y-shaped, wide, branching grooves (0.03 mm deep and 0.08 mm in diameter)
were observed near the hinge on some *Rhipidomella* hosts (Figs. 5.13, 5.14).
A few hosts were characterized by wide branching grooves along their shell
margin (Fig. 5.15a), and in some hosts these grooves were found branching from
the commissure towards the central portion of the shell (Fig. 5.16a–b). Very
distinct, deep, wide branching grooves were observed all over the shell surface of
two specimens, branching out from the hinge towards the commissure, covering
the lateral sides as well (Figs. 5.17a–b, 5.18). Another specimen also shows
characteristic evidence of prominent, deep, wide, branching grooves only near the
hinge (Fig. 5.19a–b). One specimen, partially recrystallized, has a prominent deep,
wide, slightly branching groove located centrally (Fig. 5.20). Some deep but much
narrower (0.01 mm) dendritic grooves extend from the commissure towards the
hinge of some *Rhipidomella* shells (Fig. 5.21).

Fig. 5.13 *Rhipidomella*
specimen 4. Y-shaped
branching grooves close to
the hinge (10x). *Black arrow*
points to the branching
grooves perpendicular to the
direction of rib direction

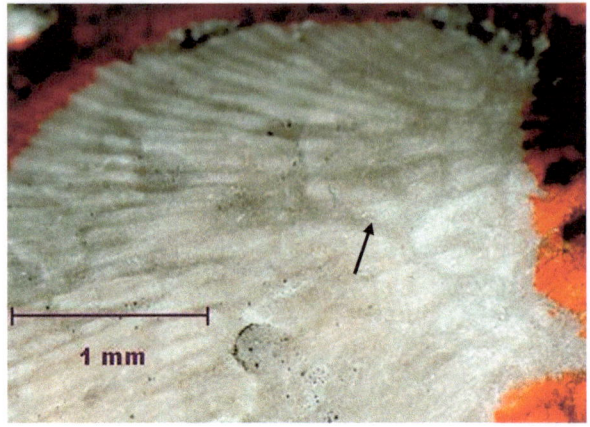

Fig. 5.14 *Rhipidomella*
specimen 2A. Deep
branching grooves close to
the hinge of the shell (16x).
Black arrow points to the
branching grooves
perpendicular to the direction
of rib direction

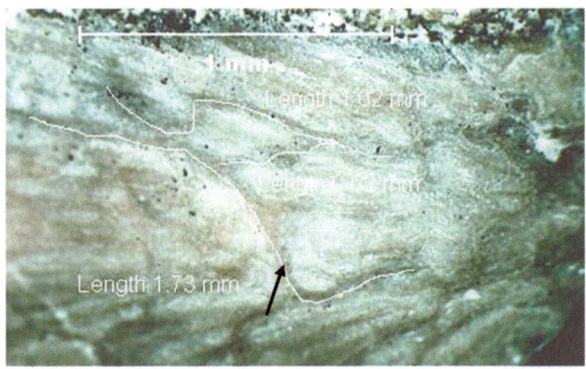

It is possible that these grooves are actually merely weathered intercostal channels and not evidence of epibiont activity. However, we see grooves crossing perpendicular to the direction of rib direction, as seen in Figs. 5.14 and 5.18, suggesting an organism actively cut through the rib.

Other specimens were characterized by three prominent sets of branching grooves parallel to each other, and quite narrow (0.03–0.04 mm), that is, half the width of the Y-shaped grooves, and shallow in nature, located centrally, extending from the shell margin towards the hinge (Fig. 5.22a–e).

5.3.4 Scars

There was evidence of two parallel, tapered scars near the commissure of a specimen that were probably a result of mechanical damage (Fig. 5.23). Two deep tapered scar marks were observed in a second specimen, one located in the central region and the other in the lower right lateral region (Fig. 5.24a–c); the lower right

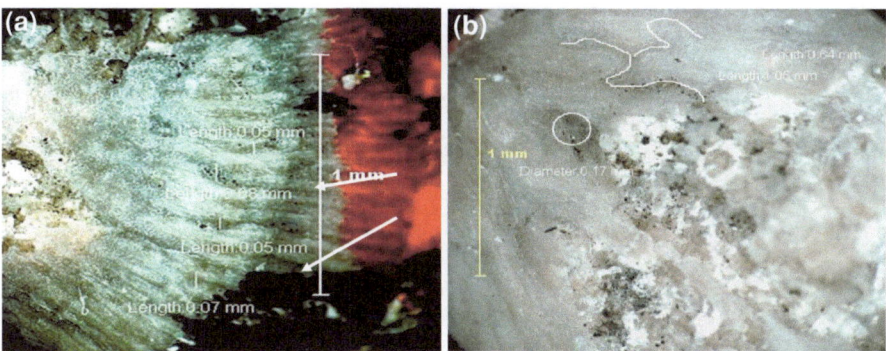

Fig. 5.15 *Rhipidomella* specimen 1A. **a** Branching grooves along the commissure of the shell (20x). *White arrows* indicate the healing of the bifurcating grooves along the anterior margin. **b** Scars and branching grooves along the hinge of the host specimen (20x)

Fig. 5.16 *Rhipidomella* specimen 17. **a** Central branching groove in a shell half-embedded in limestone matrix (8x). **b** Lower right corner of shell with some rising branching grooves (8x). *White* and *black arrow* shows signs of healing at the anterior margin

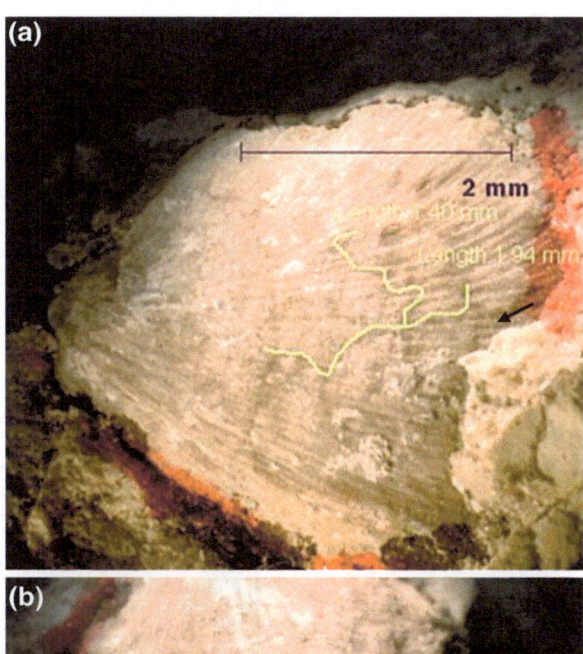

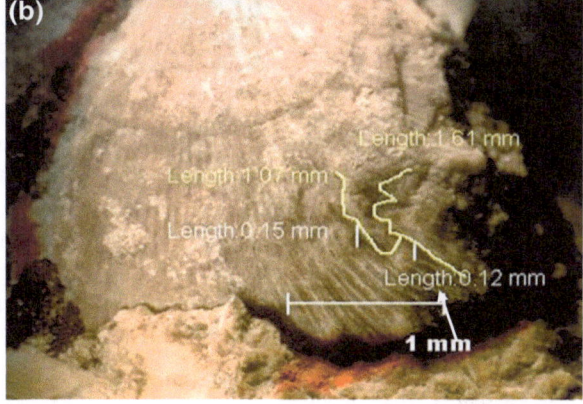

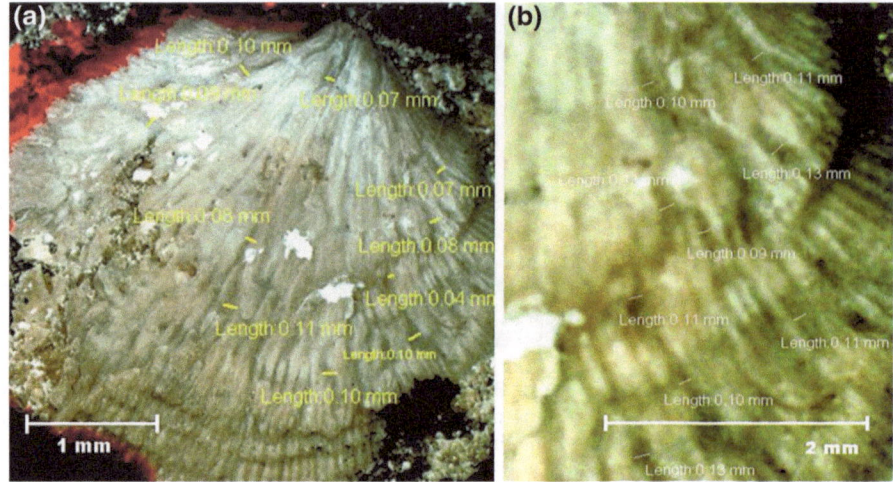

Fig. 5.17 *Rhipidomella* specimen 5. **a** Clear branching grooves all over the shell surface, branching out from the hinge towards the shell margin and along the lateral sides (8x). **b** Prominent branching grooves along the lateral sides of the shell (10x)

Fig. 5.18 *Rhipidomella* specimen 9. Distinct branching grooves all over the shell surface (8x). *Black arrow* points to the branching grooves perpendicular to the direction of rib direction

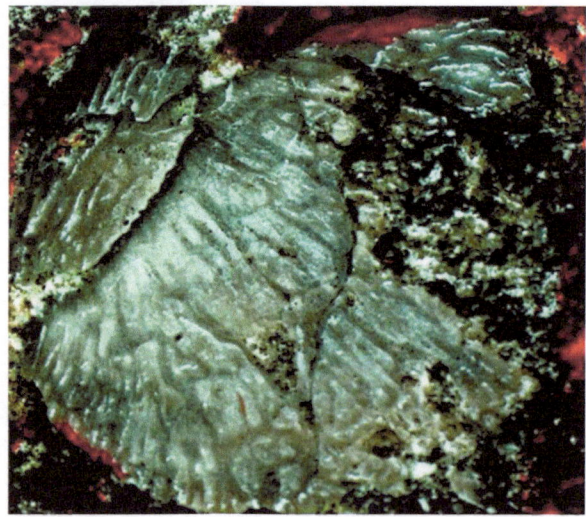

one was possibly the trace of a parasitic worm. A central deep scar mark is evident near the commissure of one specimen (Fig. 5.25a–b), which could probably have been a result of organismal interaction. A huge scar, probably a result of mechanical breakage, was observed along the broken edges of the shell margin of a specimen, with a few curved grooves, slightly bifurcating in some regions along the hinge and lateral margins of the shell (Fig. 5.26a–b). Another host specimen was characterized by two deep tapered scar marks facing each other along the

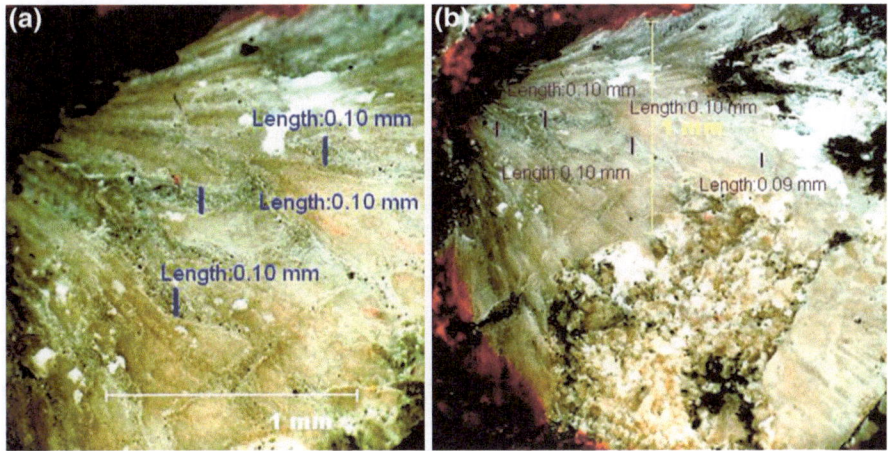

Fig. 5.19 *Rhipidomella* specimen 7 **a** Deep, wide, branching grooves close to the hinge (16x). **b** Recrystallized center and deep, branching grooves close to the hinge (10x)

Fig. 5.20 *Rhipidomella* specimen 11. Deep wide branching groove along the central portion of a partially recrystallized shell (8x). *Black arrow* shows the point where the branching groove ends, probably indicating signs of healing. (*Note* software uses the term "Length" for all linear measurements)

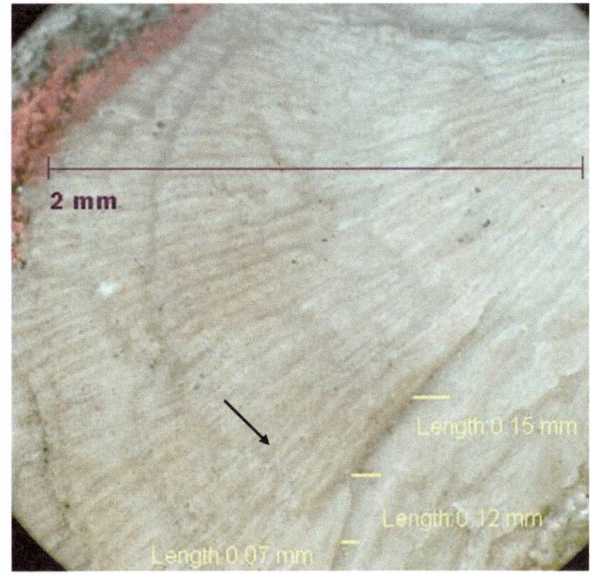

lower left lateral margin (Fig. 5.27a), similar to scar marks on *Paraspirifer* from Silica Formation (Fig. 1.1a.4). These scars were clearly produced by parasitic worms. The same host specimen shows characteristic evidence of a deep scar along the lower right lateral margin, which further implies that one side of the shell margin was compressed compared to the other (Fig. 5.27b), which was probably due to an organismal interaction.

Fig. 5.21 *Rhipidomella*
specimen 21. Deep, thin,
branching grooves extending
from the commissure towards
the hinge (16x)

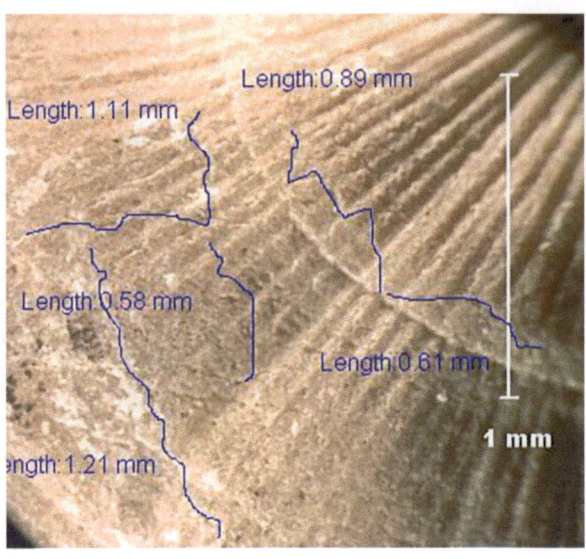

5.3.5 Large Boreholes

One *Rhipidomella* host was characterized by seven to eight large circular holes
0.16–0.28 mm in diameter located centrally, and concentrated along the grooves
of the host specimen, extending over half the shell's surface area (Fig. 5.28).

5.3.6 Mixed Traces

One specimen had a few scars (like scours) with a few branching grooves close to
the hinge region of the shell (Fig. 5.27). A large scour, probably due to mechanical
breakage, was observed along the shell margin, with a few branching grooves
spread all across the shell surface of one specimen (Fig. 5.26a–b). Elongated
straight grooves with some continuous, shallow, large holes, ranging in size from
0.1 to 1.2 mm in diameter, were observed in one *Rhipidomella* host in the central
portion of the shell, progressing towards the hinge (Fig. 5.28).

5.4 Encrustation on Non-Brachiopod Hosts

A few non-brachiopod hosts, such as a rugose coral and an internal mold of a snail,
were collected by Chris Wright from the same fossiliferous horizon of the Dundee
Formation and were examined for epibionts. The internal mold of a snail specimen, identified as *Euryzone arata*, was found encrusted with prominent, radiating,

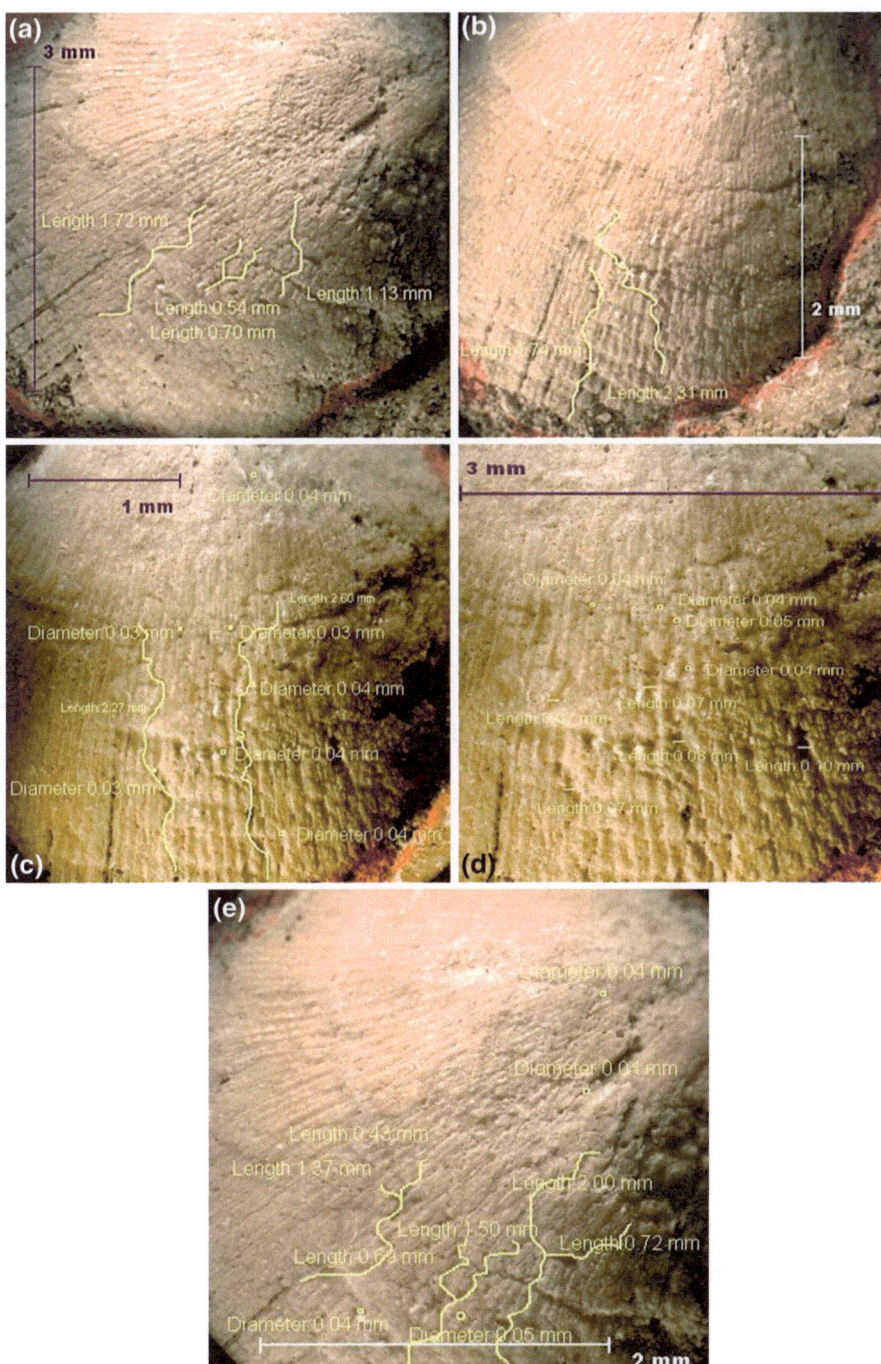

Fig. 5.22 *Rhipidomella* specimen 15. **a** Numerous branching grooves located in the central portion of the shell (8x). **b** Branching grooves progressing towards the hinge from the commissure (8x)

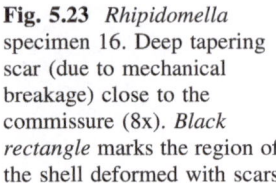

Fig. 5.23 *Rhipidomella* specimen 16. Deep tapering scar (due to mechanical breakage) close to the commissure (8x). *Black rectangle* marks the region of the shell deformed with scars

calcareous, sheet-like bryozoans (Fig. 5.29a–b). These bryozoan sheets were similar to those of the trepostome bryozoan *Leioclema* sp. that were found encrusting the host *Paraspirifer* in the overlying Silica Formation. Probably, first the bryozoans attached to the snail shell surface, after that the aperture got filled with material, and then the encrusted part of the aragonitic shell of the host dissolved away, leaving behind radiating trepostome bryozoan colonies on the internal mold of the snail.

A rugose coral specimen identified as *Zaphrenthis perovalis* collected from the Dundee Formation shows an interesting curved groove mark ending in a circular borehole, somewhat larger in size (0.24 mm in diameter) compared to most of the holes in the *Rhipidomella* specimens (Fig. 5.30a) but similar to the large boreholes shown in Fig. 5.28. This could possibly be a worm boring. Another circular trace was observed right at the top of this hole, more or less of the same diameter. Paleozoic worm borings of Devonian age are known to range from 0.05 to 0.30 mm in diameter (Cameron 1969), which suggests these holes were made by worms. Alternatively, these traces could be grooves left by a crinoid holdfast that once attached to the dead coral lying at the seafloor bottom (Fig. 5.30b). However, lack of evidence of traces of crinoid columnals along the groove suggests this was a worm boring.

These specimens suggest that the environment in the Dundee was favorable for encrusting organisms, but yet epibionts were rare on the brachiopod hosts.

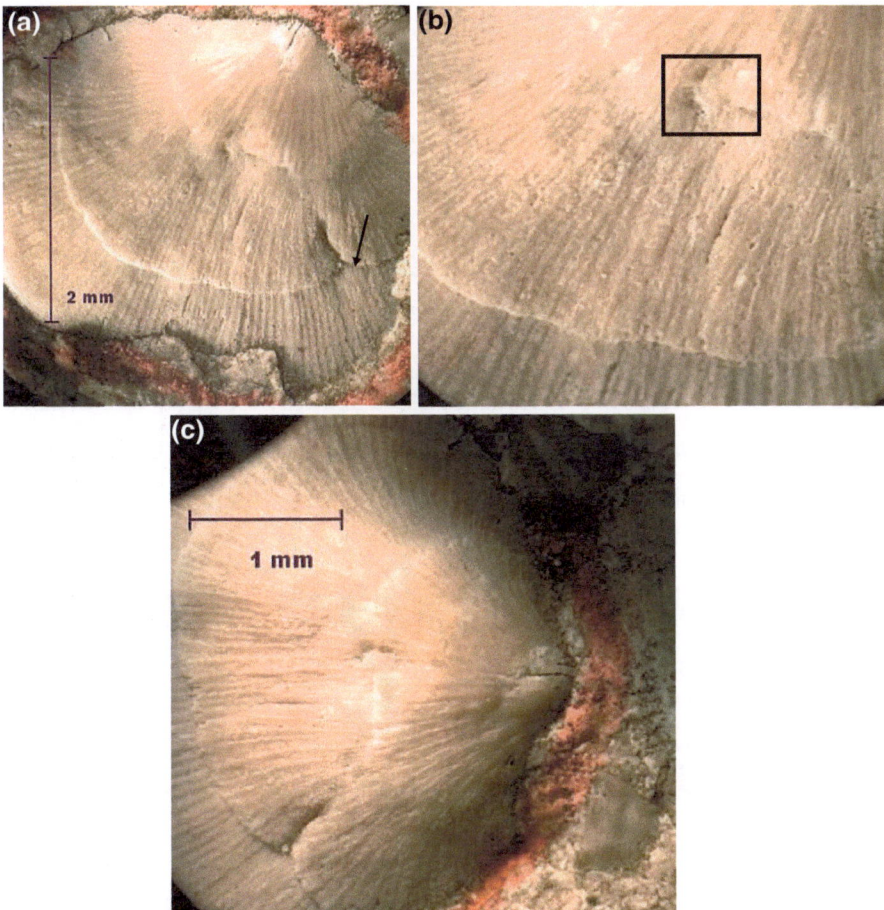

Fig. 5.24 *Rhipidomella* specimen 22. **a** Two deep scars, one in the central region close to the hinge, the other on the right lateral side; tiny punctae close to the margin (8x). *Arrow* shows the scar mark. **b** Central deep scar with small punctae (12.5x). *Square* shows the scar healing region. **c** Central and right lateral scar (8x)

5.5 Identification of Trace-Makers

One *Rhipidomella* specimen has very distinct, widespread, deep and wide branching grooves all along the shell surface (Fig. 5.23a). This dendritic pattern observed on the host shell surface is characteristic (in size and shape) of traces of soft-bodied bryozoans of order Ctenostomata belonging to class Gymnolaemata. These ctenostome bryozoans are uncalcified forms that bore into calcareous substrates. Traces of such grooves (0.15–0.40 mm) are of a similar diameter to those of boring forms known from Ordovician to Recent age, and hence these branching grooves on the Devonian brachiopods are likely the product of ctenostome

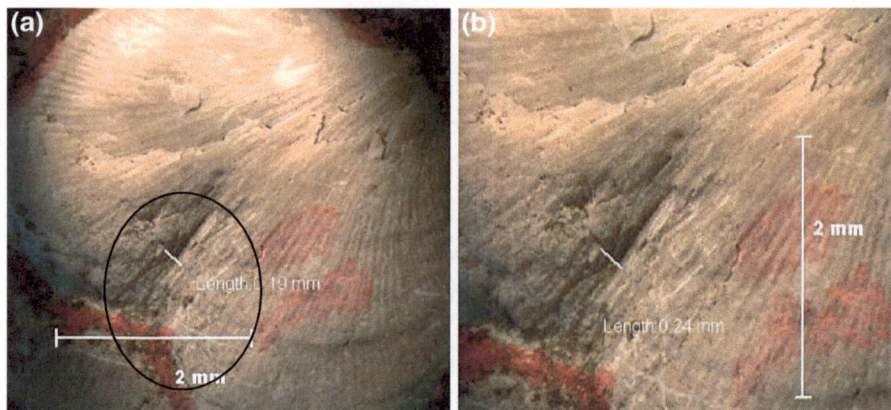

Fig. 5.25 *Rhipidomella* specimen 24. **a** Whole shell with a deep central groove and some punctae near shell margin (8x). *Black circle* marks the region of the deformed shell with a scar. **b** Central deep scar with angular holes (8x)

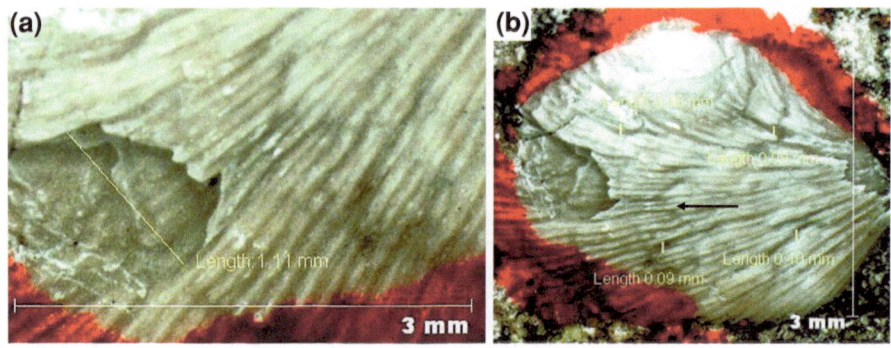

Fig. 5.26 *Rhipidomella* specimen 10. **a** A deep, large scar on a shell surface along the broken edges combined with some V-shaped grooves (10x). **b** Numerous branching grooves on the shell surface with a lower big scar (8x). *Black arrow* marks the V-shaped groove

bryozoans. Moreover, Gluchowski (2005) described the presence of ctenostome bryozoans on Upper Eifelian-aged crinoid columnals collected from the Devonian strata in the lower part of the Skaly Beds in the Holy Cross Mountains, Poland (Fig. 5.31), which further supports their identification here.

Skeletal bryozoans were not observed encrusting on any brachiopod shell, which could mean they were not present in this environment or they were not preserved for some reason. However, an internal mold of a gastropod from the same unit was found encrusted with a calcareous, radiating sheet-like bryozoan (Fig. 5.29a). Brachiopods (specifically, *Paraspirifer*) from the overlying Silica Shale were heavily encrusted by a sheet-like cover of similar colonies of

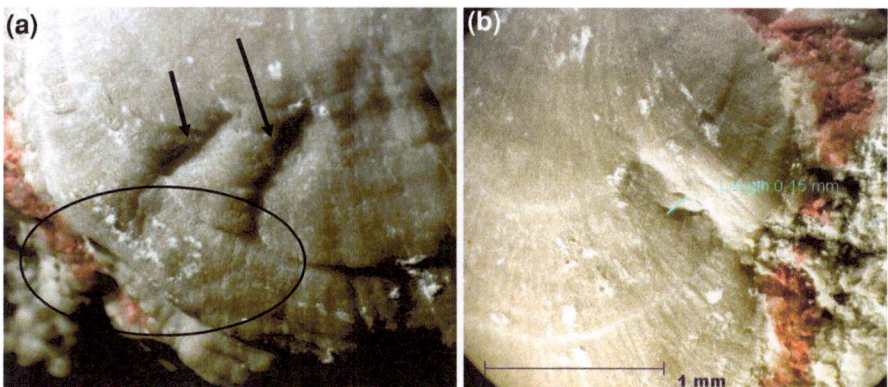

Fig. 5.27 *Rhipidomella* specimen 13 **a** Two deep scars along the left lateral margin (12.5x) similar to the scars on *Paraspirifer* collected from Silica as shown in Fig. 1.1a.4. *Black arrows* point to the two parallel deep scars. *Black circle* marks the healing of the scars. **b** A deep scar on the right lateral margin, compressing one side of the shell surface with respect to the other side (12.5x)

Fig. 5.28 *Rhipidomella* specimen 3. Central grooves branching out from the hinge with shallow larger holes in the central portion of the shell (8x)

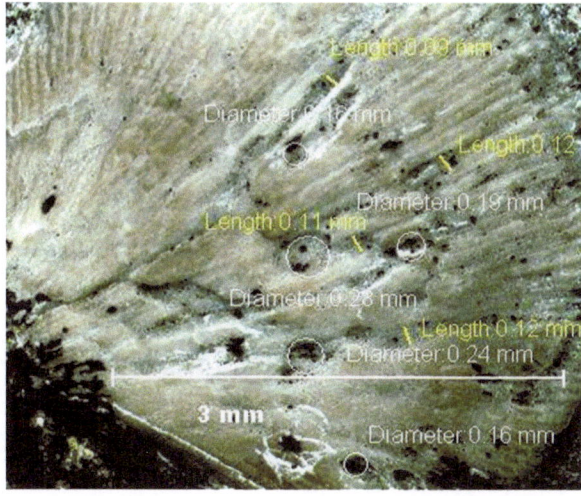

trepostome bryozoans, *Leioclema* sp., as illustrated by Sparks and others (1980, Pl. 18. Figs. 1.2, 3.3, 3.4). Therefore, the sheet-like bryozoan colony on the gastropod was identified as a *Leioclema* species (Fig. 5.29b) belonging to the order Trepostomata. This suggests that there were sheet-like bryozoans present in the Dundee environment. Yet this type of bryozoan did not encrust any Devonian brachiopod host from the Dundee Formation. It is not clear why they avoided encrusting brachiopods and preferred snails. Only a single brachiopod shell was observed with a very flat sheet-like encrustation, very different from the snail encruster in being extremely thin and having very minute pores. This was

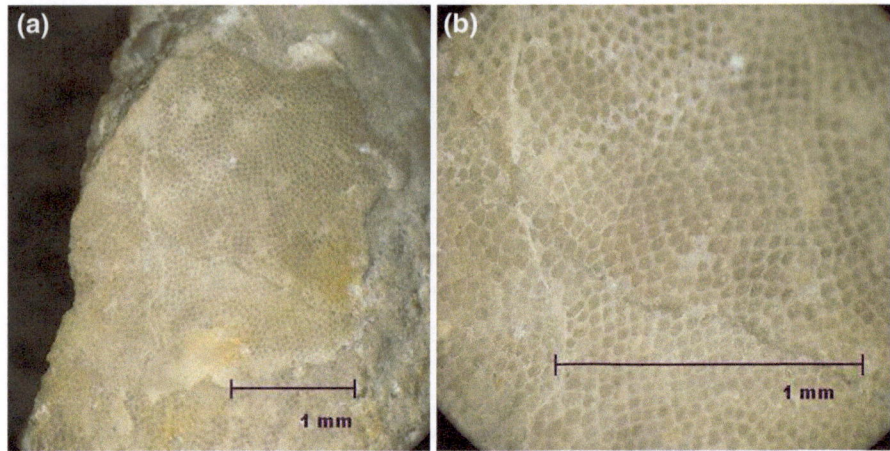

Fig. 5.29 *Euryzone arata* specimen 1. **a** An internal mold of gastropod with sheet-like calcareous encrustation of a trepostome bryozoan *Leioclema* (8x). **b** Close up of *Leioclema* colony (20x)

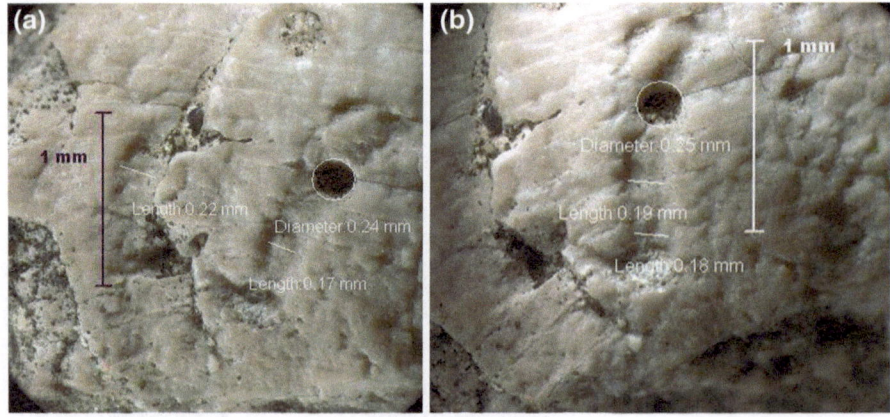

Fig. 5.30 *Zaphrenthis perovalis*. **a** A *circular* hole at the end of a continuous, wide, curved groove present close to the upper large margin of a horn coral (12.5x). *Arrow* shows another *circular* trace. **b** Trace of another *circular* hole, possibly a worm boring, above the borehole on the same coral specimen (12.5x)

identified as a bryozoan encrustation (Fig. 5.18), but it was impossible to determine the type of bryozoan encruster in this particular specimen.

Scar marks determined in a few specimens were mostly made by some organisms while the brachiopods were growing, leading to irregular growth of the shell or causing deformation of the shell; only a few were caused by mechanical

Fig. 5.31 Ctenostome
bryozoans encrusting entire
crinoid columnals collected
from the Upper Eifelian
Skaly Beds in Skaly Village,
Holy Cross Mountains in
Poland (from Gluchowski
2005)

breakage. Two prominent scars (Figs. 5.24a, 5.27a) were very similar to the
damage present on the brachiopod shell collected from the Silica Formation
(Fig. 1.1a.4). These scar producers were identified as parasitic worms. The cause
of scar marks present in other shells (Figs. 5.23, 5.25) is uncertain. These could be
a result of mechanical breakage. Scar marks observed in two other specimens were
probably produced by some unknown organisms (Figs. 5.26, 5.27b).

Large circular boreholes of diameter 0.16–0.28 mm were observed on a
Rhipidomella shell (Fig. 5.28) and two holes of diameter 0.24 mm were found on a
rugose coral specimen collected from the same fossiliferous horizon of the Dundee
Formation. Paleozoic worm borings on brachiopod shells and corals were found to
range in diameter from 0.05 to 0.3 mm, thus supporting the identification of these
traces as worm borings (Cameron 1969).

References

Williams A, Brunton CHC, Carlson SJ (1997). Treatise on Invertebrate Paleontology, part H,
     Brachiopoda (revised) (Kaesler RL, ed). Geological Society of America and The University of
     Kansas, Boulder, Colorado and Lawrence, Kansas, 539 p
Rowell AJ, Grant RE (1987) Phylum Brachiopoda, pp 463–488. In: Boardman RC, Cheetham
     AH, Rowell AI (eds) Fossil invertebrates. Blackwell Science, Cambridge, p 713

Clarkson ENK (1986) Invertebrate paleontology and evolution, 2nd edn. Allen and Unwin,
    London 382 p
Cameron B (1969) Paleozoic shell boring annelids and their trace fossils. Zoologist 9:689–703
Gluchowski E (2005) Epibionts on upper Eifelian crinoid columnals from the Holy Cross
    Mountains, Poland. Acta Palaeontol Polonica 50(2):315–328

Chapter 6
Discussion

6.1 Epizoan–Host Relationships

Rhipidomella shells were characterized by different biological activities. Other potential brachiopod hosts, *Strophodonta*, *Atrypa*, *Rhynchotrema*, and *Mucrospirifer*, show no signature of any biological activity.

6.1.1 Sheet-Like Spotty Encrustations

The flat sheet-like encrustation (Fig. 5.12) by an undetermined group of bryozoans underneath the broken shell surface suggests postmortem encrustation. Probably, the exposed shell surface got fragmented or its upper sheet got flipped off by high-energy storm currents and then encrustation took place underneath the cover. The host was long dead, thus there were no biotic interactions between the epibiont and the host. The epibiont probably took advantage of the hard substrate and grew on it.

6.1.2 Branching Grooves

Brachiopods, in general, take in food through inhalant currents near the hinge and then give out the extra food and waste by exhalant currents near the central commissure (Fig. 6.1). Two host specimens with deep branching grooves ranging from 0.05 to 0.08 mm in diameter along the anterior margin of the shell (Figs. 5.12 a, 5.16a, b) suggest the symbiotic ctenostomate bryozoan epibionts possibly benefited from the host exhalant currents. Therefore, the position of the trace left by bryozoans firmly suggests that their relation with the host was commensal. These show evidence of healing of these grooves along the anterior margin, which suggests this biological activity took place during the host's life.

R. Bose, *Devonian Paleoenvironments of Ohio*, SpringerBriefs in Earth Sciences, 49
DOI: 10.1007/978-3-642-34854-9_6, © The Author(s) 2013

Deep branching grooves along the hinge area of some host specimens (Figs. 5.13, 5.14, 5.15b) suggest ctenostomate bryozoans were probably benefiting from host inhalant currents, thereby harming the host by directly taking away food from the brachiopod. This suggests a parasitic relation between the host and the ctenostomate bryozoans.

Some deeply penetrating branching grooves 0.04–0.11 mm in diameter, extending from the commissure and lateral margins and then converging towards the hinge (Fig. 5.17a, b), suggest these soft ctenostomate bryozoans attached to the shell were benefiting from strong exhalant currents of the host. The position of the epibionts suggests they had a commensal relation with the host, thereby proving that the host was living during the time of encrustation.

One host specimen with a somewhat flat surface was characterized by straight and branching grooves converging from the commissure towards the hinge (branching out from the hinge) (Fig. 5.18). Part of the shell was recrystallized and hence it was hard to tell the time of encrustation. One particular specimen (Fig. 5.19a, b) with branching grooves 0.10 mm in diameter along the hinge region shows evidence of breakage and postburial recrystallization. Hence, it is hard to tell if any relationship existed between the host and the epibiont. Soft, branching bryozoans were attached to the host, possibly for protection. A partially recrystallized host specimen (Fig. 5.20) characterized by a deep, straight groove of diameter 0.07–0.15 mm along the central area, branching at the lower end towards the commissure, is also recognized as the biological trace of a soft bryozoan. It seems that this groove was later healed along the shell margin, which further suggests the shell was alive during this activity. Deep, narrow criss-cross branches with small punctae along the major growth lines extending from commissure to hinge suggest epibiont infestation at a particular time. This branching or crossing pattern was different from the ctenostomate bryozoan traces. It seems in these specimens (Fig. 5.21) that one boring trace crosses another one. Borings seemed straight, cutting across the radial ornamentation in some cases, while, in other cases, generally approximating the orientation of the major ornamentation of the

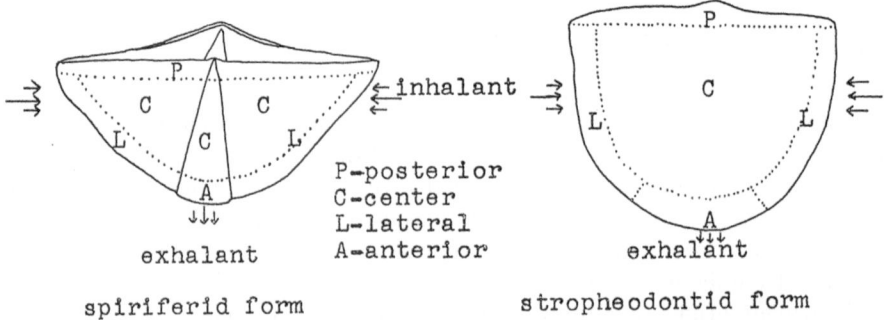

Fig. 6.1 Designated positions for epifauna determined by inhalant and exhalant currents (from Steller 1965)

host. These traces were very similar to the trace fossil *Vermiforichnus clarki,* very common on Middle Devonian brachiopod specimens of the Silica Formation (Hoare and Walden 1983). These borings along the commissure probably caused cessation of the growth activity of the host. Deformation of the shell along the anterior edge to some extent could be due to the detrimental effect of the worm borer on the host specimen, thereby ceasing its growth. These activities were taking place during the life of the host. Numerous branching grooves progressing from the commissure towards the hinge are predominant in the antero-central region of the host specimen (Fig. 5.22a–e), and suggest these ctenostomate bryozoans also probably benefited from the host feeding exhalant currents. Hence, this host-epibiont relationship is an example of commensalism, in which the epibiont is benefited without affecting the host. Thereby, the trace making was clearly during its life. The host was deformed close to the hinge region; however, this was probably due to fragmentation or abrasion caused by storm action. Some straight and slightly curved grooves of diameter 0.09–0.10 mm along the central region of the host suggest the activity of some ctenostomate bryozoans over the host specimen (Fig. 5.26a, b). These grooves, progressing towards the hinge and following the radial ornamentation, suggest some soft bryozoans, seeking shelter in the hard substrate, left these traces. The shell was broken near the anterior margin; probably it underwent mechanical post-burial damage and, hence, it was difficult to tell the time of this biological activity.

Gluchowski (2005) discovered five crinoid species from the Upper Eifelian of Poland infested by ctenostome bryozoans, entirely soft-bodied organisms (Fig. 6.1), and skeletal epibionts. He found that all these epibionts that settled on the living crinoid hosts were not detrimental to the host and that the biotic interaction was best described as commensalism. This further supports the presence of ctenostome bryozoans on brachiopods of the Upper Eifelian-aged Middle Devonian Dundee Formation, and their relationship with the host being commensal.

6.1.3 Scars

There were inconclusive scar marks on specimens (Figs. 5.23, 5.26a, b) interpreted as caused by mechanical breakage and not by any organismal interaction, as evidenced from shell abrasion and fragmentation. There were no signs of healing from inside the shell, which suggests the shell was dead and that the dead host was fragmented due to storm activity. Two deep, tapered scars (Fig. 5.24a–c), one located in the central region and the other along the right lateral region of the host, also shed some light on the host-epibiont relationship. The central scar, however, suggests deformation of the shell and ceasing of the growth activity of the host for some time. The healing of the scar implies that this biotic interaction occurred during its life. The lower, right lateral scar was deep, and signs of healing by the secretion of the host (Fig. 5.24c) on the younger successive layer suggest biological activity during its life. Such a deep groove along the lower right lateral region makes it clear that the host was being attacked very energetically, and hence a parasitic relation was deduced from the undetermined epibiont's activity. There were no signs of abrasion or fragmentation or any mechanical postburial damage, which suggests this scar mark or damage on the host shell surface was not due to storm activity. Another specimen mainly characterized by two deep, parallel, tapered scar marks located near the lower left lateral margin (Fig. 5.27a) shows some sign of scar healing along the anteriormost edge. This suggests the shell was attacked during its life time. There were similar scar marks on a brachiopod host collected from the Silica Formation (Fig. 1.1a), which suggest that these scar marks were produced by parasitic worms. The position of the scar also suggests a parasitic relationship between the two. Possibly, the epibiont was trying to feed on the soft tissues of the host. Another specimen (Fig. 5.27b) deformed along the lower, right lateral region of the host shows a deep, tapered scar that does not show healing along the margin, suggesting postmortem damage.

A somewhat deformed host specimen (Figs. 5.26a, b) is characterized by a deep, large scar (like a scour mark) with no sign of healing, suggesting postmortem mechanical breakage. This deformation of the host could also possibly be due to postburial diagenesis.

6.1.4 Large Borehole

A single host specimen, marked by wide, straight grooves stopping in large boreholes along the central region (Fig. 5.28) and then progressing towards the hinge, suggests that some epibiont was boring through the shell surface. Holes part way through the shell in general suggest epibiont activity. Close to the hinge, the groove bifurcates, passing underneath the shell cover and further continuing until it reaches the hinge. This grooving underneath the external shell surface suggests postmortem boring. The shell must have been fragmented and abraded by some strong storm action and then bored. These branching grooves could be mistaken for part of the mantle canal system typical of many brachiopods. However, the diameter of the canals is typically less than 0.001 mm in diameter in orthid brachiopods, while the diameter of the grooves ranges from 0.15 to 0.40 mm. Hence, these traces appear to be epibionts, specifically, worm borings, and not part of the microstructure of the brachiopod shell itself.

6.1.5 Trace-Marks on Non-Brachiopod Hosts

The internal mold of a gastropod found in the Dundee Formation was found encrusted by the radiating sheet-like calcareous bryozoan *Leioclema,* which was not seen encrusting any brachiopod host. These types of epibionts are more common in siliciclastic muds (Sparks et al. 1980). It is possible that this gastropod specimen was transported from deep water by strong currents due to storm activity and was not co-occurring in life with the *Rhipidomella* brachiopods. The rugose coral specimen also possibly might have been transported from a deeper substrate, where it was bored and grooved. If these were not traces of a worm borer on a coral but the traces of a crinoid holdfast, then it is possible that the host was encrusted when it was lying at the bottom with its upper margin exposed. The crinoid holdfast probably got attached to it and was then growing on the dead host. However, the diameter of the borehole opening was 0.24 mm. Paleozoic worm borings with diameters ranging from 0.01 to 0.3 mm are known from the prior work of researchers (Cameron 1969). Hence, these circular borehole openings were more likely the traces of worms and not crinoid holdfasts attached to them. These worms probably just took advantage of the hard substrate. This was mere substrate boring and, hence, implies no specific relationship between the host and the borer.

6.2 Four Possible Causes for Rare Encrustation in the Dundee Formation

Firstly, it is known that the carbonate environment in general is characterized by high levels of abrasion (Wilson and Taylor 1999). Skeletal bryozoans were absent on the *Rhipidomella* hosts of the Dundee Formation, possibly due to high levels of abrasion, possibly related to strong storm action. After burial, possibly many brachiopod shells in the Dundee got deformed and then they were subjected to postburial diagenesis. Hence, a lot of specimens show evidence of mechanical postburial diagenesis, thus masking epibionts. Only a few soft uncalcified encrusting bryozoans were observed on the host specimens. However, a non-brachiopod host, *Euryzone arata,* has been found from the same unit with a sheet-like calcified epibiont encrusted on it, which suggests that there were skeletal bryozoans present in the Dundee environment but they were not encrusting the brachiopods. However, it is possible that the gastropod was transported from a deeper water environment and was not actually living in the same community as the brachiopods. In brief, the lack of encrustation could be due to poor preservation.

Secondly, the portion of the Dundee Formation studied was not very extensive. It is therefore difficult to draw conclusions about the frequency of encrustation for all the brachiopods of the Dundee Limestone. Study of the frequency of encrustation on the brachiopods of the Dundee Limestone sampled from a more extensive area, particularly from non-storm bed horizons, would be useful to determine the cause of rare encrustation. For instance, it would be worthwhile to study the same unit exposed in other quarries. Brachiopods in other limestones from other localities have been reported to have been encrusted by skeletal organisms. Hence, the apparent lack of encrustation could be due to incomplete sampling of the Dundee Formation.

Thirdly, the cause for this rare encrustation on the brachiopod specimens possibly relates to the rate of bioturbation in the Dundee Formation. Fieldwork has determined that much of the Dundee was intensely bioturbated, implying a soft substrate (Figs. 5.7, 5.8, 5.9, 5.10). Therefore, it is highly possible that bioturbation mixed shells down into the soft substrate before epibionts could attach. Hence, the lack of encrustation in the Dundee Formation is real, and caused by biological activity keeping the substrate well-mixed and reducing the residence time of shells on the seafloor.

Fourthly, the Dundee Formation may have had low nutrient levels, as indicated by the presence of abundant corals and worm borers. Thus, the Dundee was a mesotrophic site and the low amount of encrustation could be directly due to low productivity.

I strongly think that the third cause, that is, intense bioturbation, is a more likely explanation for the low amount of encrustation in the Dundee Formation, while at the same time, the first cause, that is, poor preservation of the shells cannot be wholly neglected. Therefore, future sampling of non-storm beds in eutrophic sites in an extensive area could possibly rule out the first, second and third causes and further confirm my hypothesis for the rarity of encrustation in the Dundee.

References

Cameron B (1969) Paleozoic shell boring annelids and their trace fossils. Zoologist 9:689–703
Gluchowski E (2005) Epibionts on upper Eifelian crinoid columnals from the Holy Cross Mountains, Poland. Acta Palaeontol Pol 50(2):315–328
Hoare RD, Walden RL (1983) Vermiforichnus (Polychaeta) borings in Paraspirifer bowneckeri (Brachiopoda: Devonian). Ohio J Sci 83(3):114–119
Sparks DK, Hoare RD, Kesling RV (1980) Epizoans on the brachiopod *Paraspirifer bowneckeri* (Stewart) from the Middle Devonian of Ohio. University of Michigan Museum of Paleontology. Papers on Paleontology, 23:1–50
Steller DL (1965) The epifaunal elements on the Brachiopoda of the silica formation. Unpublished master's thesis, Bowling Green State University, Bowling Green, p 76
Wilson MA, Taylor PD (1999) Carbonate hardgrounds and their faunas in a coarse siliciclastic environment: the Qahlah Formation (Upper Cretaceous) of the Oman Mountains. Geol Soc Am Abstr Programs 31(7):105

Chapter 7
Conclusion

A detailed study of brachiopods collected from the Devonian Dundee Formation of Ohio for epibionts gives a fair sense of the amount of encrustation in this unit. From previous literature, it is already known that the strophodonts and spiriferids from the overlying siliciclastic unit, the Silica Formation, were heavily encrusted with many skeletal organisms known to have attached to hard substrates in this unit. The present study of the Dundee Formation has clearly shown that strophodonts and spiriferids show no evidence of encrustation. Instead, a few *Rhipidomella* shells show some evidence of different types of biological activities. These *Rhipidomella* shells show evidence of branching grooves, scars, sheet-like encrustation, and large borehole openings. The results of this study clearly demonstrate that only this type of brachiopod was encrusted, mainly by soft bryozoans, and there is rare evidence of skeletal organisms attaching to the hard substrates in this carbonate unit. The overall amount of encrustation in the Dundee Formation (carbonate) compared to the overlying Silica Formation (siliciclastic) has been found to be very low. Sparks et al. (1980) found 582 out of 586 specimens of *Paraspirifer bownockeri* from the Middle Devonian Silica Formation of Ohio encrusted. That is around 99 % of brachiopod shells encrusted in the overlying siliciclastic unit based on their study, while from this present study of the underlying Dundee Formation, eight spiriferids collected showed no epibionts. However, 21 out of 245 specimens of brachiopod shells showed evidence of ecological interactions, that is, only 8.6 % of the shells were encrusted. No study has yet been performed on the *Rhipidomella* specimens of the Silica Shale. Therefore, a detailed, comparative, quantitative analysis of epibionts on both spiriferids and rhipidomellids from both the siliciclastic and carbonate units needs to be performed in future in order to get a clear sense of the low frequency of encrustation in this carbonate unit compared to the siliciclastic unit, based on the study of the same brachiopods. Further, it is also important to determine the environment of deposition for brachiopods encrusted in other Devonian limestones.

This research has contributed a large field collection from the Dundee Formation, which will be useful in documenting the fossil content of this unit for future workers. Furthermore, this study has increased our understanding of epibiont-host relationships. I believe that this research will also assist future workers to compare the encrustation patterns on brachiopod hosts of the Dundee Limestone with that of other Devonian brachiopods, from both carbonate and siliciclastic settings.

Reference

Sparks DK, Hoare RD, Kesling RV (1980) Epizoans on the brachiopod *Paraspirifer bowneckeri* (Stewart) from the Middle Devonian of Ohio. University of Michigan Museum of Paleontology. Papers on Paleontology, No. 23:1–50

About the Author

Dr. Rituparna Bose obtained her Ph.D. from Indiana University, Bloomington. For her outstanding Ph.D. dissertation she was rewarded with a Springer Theses Prize. She is currently an adjunct Assistant Professor at the City University of New York and has been interviewed as an expert in the field of biodiversity by the Times of India (leading news daily in her home country). She has won numerous awards in her career including the nationally recognized Theodore Roosevelt Memorial Grant awarded by the American Museum of Natural History.

She has been recently invited to serve as an Editor for Acta Paleontologica Sinica by the Chinese Academy of Sciences. She is the Associate Editor-in-Chief for the International Journal of Environmental Protection and the Associate Editor for the Journal of Geography and Geology at the Canadian Center of Science. She also serves on the editorial board of some of the most prestigious journals in geology including Historical Biology: An International Journal of Palaeobiology (Taylor and Francis), Bulletins of American Paleontology (Paleontological Research Institute, Cornell University), Springer Plus (Earth and Environmental Science) and Geological Journal (Wiley).

The foreword of this book has been written by Prof. David Harper who is presently the Vice President of the Palaeontographical Society and Chairman of the International Subcommission on Ordovician Stratigraphy. He was previously the President of the International Palaeontological Association (2006–2010). He is the Editor of the journal Lethaia and Advisory Editor of the Journal of the Geological Society, London.

R. Bose, *Devonian Paleoenvironments of Ohio*, SpringerBriefs in Earth Sciences,
DOI: 10.1007/978-3-642-34854-9, © The Author(s) 2013